S0-DVD-121

AN INSIDE LOOK

DISCOVERING THE UNIVERSE

For a free color catalog describing Gareth Stevens Publishing's list of high-quality books and multimedia programs, call 1-800-542-2595 (USA) or 1-800-461-9120 (Canada). Gareth Stevens Publishing's Fax: (414) 332-3567.

Gareth Stevens Publishing would like to thank noted science author Greg Walz-Chojnacki of Milwaukee, Wisconsin, for his kind and professional help with the information in this book. Mr. Walz-Chojnacki is the author of *Celestial Delights: The Best Astronomical Events Through 2001* and *Comet: The Story Behind Halley's Comet.*

Library of Congress Cataloging-in-Publication Data available upon request from publisher. Fax: (414) 332-3567 for the attention of the Publishing Records Department.

ISBN 0-8368-2724-4

This North American edition first published in 2000 by
Gareth Stevens Publishing
A World Almanac Education Group Company
330 West Olive Street, Suite 100
Milwaukee, WI 53212 USA

Illustrators: Jeff Bowles, David Hardy, Sebastian Quigley, Mike Saunders, Steve Weston, and Gerald Witcomb
Gareth Stevens editors: Christy Steele and Heidi Sjostrom

Printed in Mexico

1 2 3 4 5 6 7 8 9 04 03 02 01 00

AN INSIDE LOOK

DISCOVERING THE UNIVERSE

Stuart Clark

Gareth Stevens Publishing
A WORLD ALMANAC EDUCATION GROUP COMPANY

AN INSIDE LOOK

CONTENTS

Big Bang

Most astronomers believe the Universe began about 15 billion years ago with an event known as the Big Bang. The Universe was a tiny point of very hot energy before the Big Bang. After the Big Bang, this tiny point suddenly expanded. It took only several seconds for the entire Universe, and all the material within it, to grow. Within the first few minutes, enough matter was created to form everything in space. Even the particles that make up this book were formed in the first few seconds after the Big Bang.

The Universe began in a very tiny space. All the matter in the Universe was pressed tightly together. Then atoms began to form and expand this dense material. Atoms are the building blocks of matter. Over time, the Universe continued to expand. It became less dense as it grew. Today, space is mostly empty. Dense areas filled with planets, stars, and galaxies are scattered throughout it.

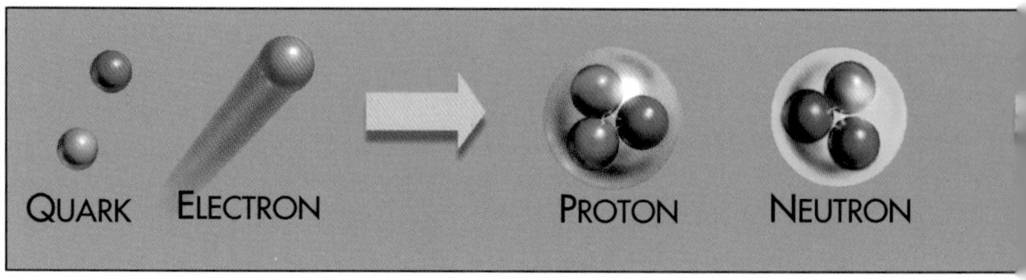

QUARK
ELECTRON
PROTON
NEUTRON

Formation of matter
A millionth of a second after the Big Bang, simple particles called quarks and electrons appeared. Quarks then joined to form neutrons and protons. A single proton is known as a hydrogen nucleus. Later, neutrons and protons joined to make helium nuclei. Finally, electrons began to orbit these nuclei, thus forming atoms.

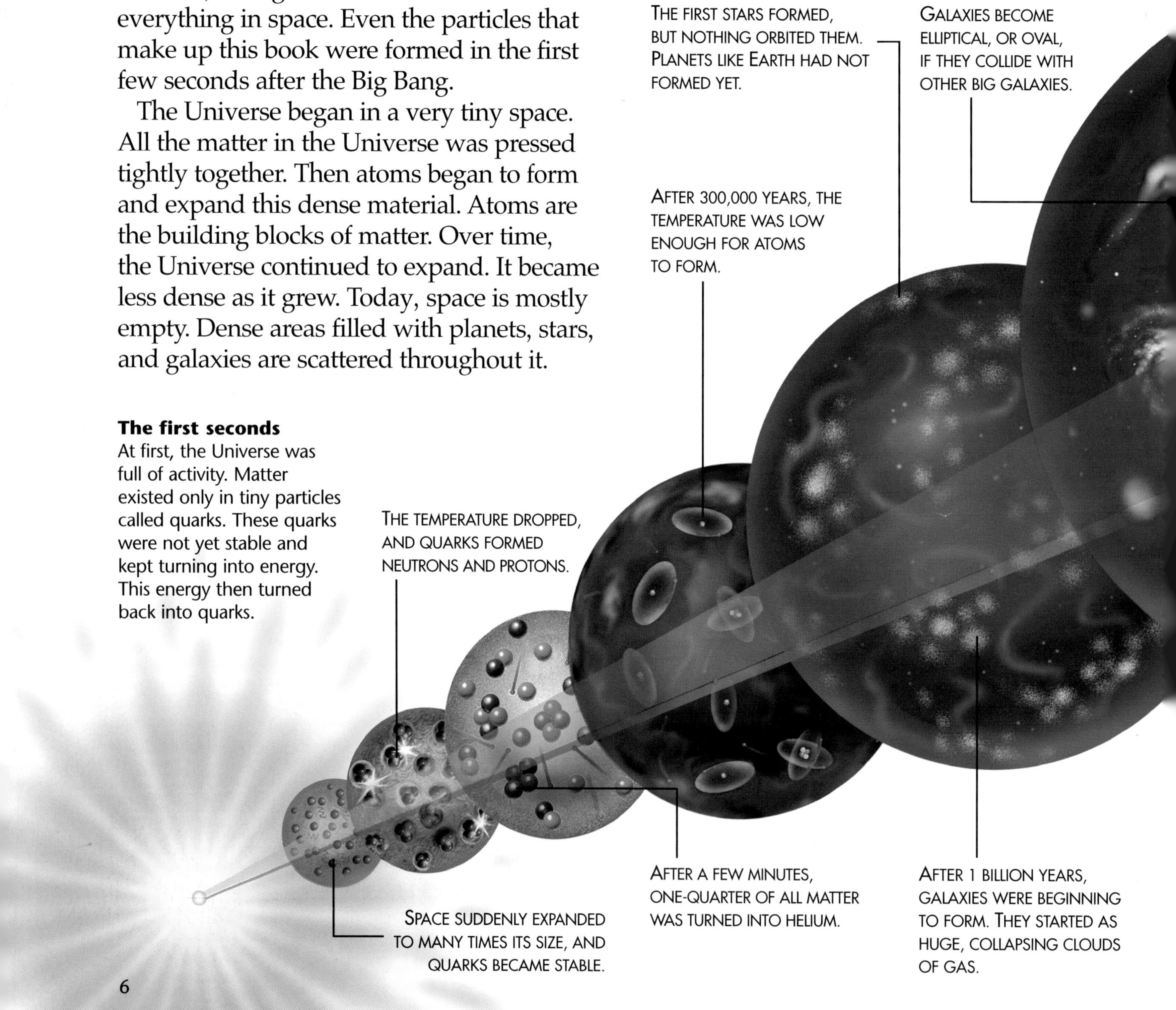

The first seconds
At first, the Universe was full of activity. Matter existed only in tiny particles called quarks. These quarks were not yet stable and kept turning into energy. This energy then turned back into quarks.

THE FIRST STARS FORMED, BUT NOTHING ORBITED THEM. PLANETS LIKE EARTH HAD NOT FORMED YET.
GALAXIES BECOME ELLIPTICAL, OR OVAL, IF THEY COLLIDE WITH OTHER BIG GALAXIES.
AFTER 300,000 YEARS, THE TEMPERATURE WAS LOW ENOUGH FOR ATOMS TO FORM.
THE TEMPERATURE DROPPED, AND QUARKS FORMED NEUTRONS AND PROTONS.
AFTER A FEW MINUTES, ONE-QUARTER OF ALL MATTER WAS TURNED INTO HELIUM.
AFTER 1 BILLION YEARS, GALAXIES WERE BEGINNING TO FORM. THEY STARTED AS HUGE, COLLAPSING CLOUDS OF GAS.
SPACE SUDDENLY EXPANDED TO MANY TIMES ITS SIZE, AND QUARKS BECAME STABLE.

HYDROGEN NUCLEUS
HELIUM NUCLEUS
HYDROGEN ATOM
HELIUM ATOM
PLANETS LIKE EARTH FORMED AROUND STARS.
TEN BILLION YEARS AFTER THE BIG BANG, THE SUN AND THE PLANETS IN OUR SOLAR SYSTEM FORMED.
OUR SOLAR SYSTEM IS IN THE MILKY WAY GALAXY.
THE UNIVERSE IS NOW ABOUT 15 BILLION YEARS OLD AND CONTINUES TO EXPAND.
1
2
3

Birth of a solar system
Solar systems like ours formed when gas clouds in galaxies collapsed. First, a gas cloud began to shrink, and a star began to form at its center (1). As the cloud collapsed more, gases and dust formed a disk around the star (2).

Dust in this disk collided and stuck together, gradually turning into larger dust grains. For millions of years, the dust grains continued to grow until they formed planets. The planets began to orbit the star, creating a solar system (3).

The Universe

The Universe is so large that astronomers measure most distances within it in light-years. One light-year is the distance light travels in a year. This equals 5.9 trillion miles (9.5 trillion kilometers). A ray of light travels 186,291 miles (299,792 km) per second, faster than anything else in the Universe.

Many different types of objects exist in the Universe. Some objects are very large, while others are tiny. The largest objects in the Universe are gigantic filaments. Filaments consist of many superclusters, and superclusters are huge groups of galaxies. Galaxies are made of millions of stars, along with the planets, moons, and other objects that orbit the stars. Some galaxies are still forming stars today, but others stopped millions of years ago. The smallest solid objects in the Universe are dust particles.

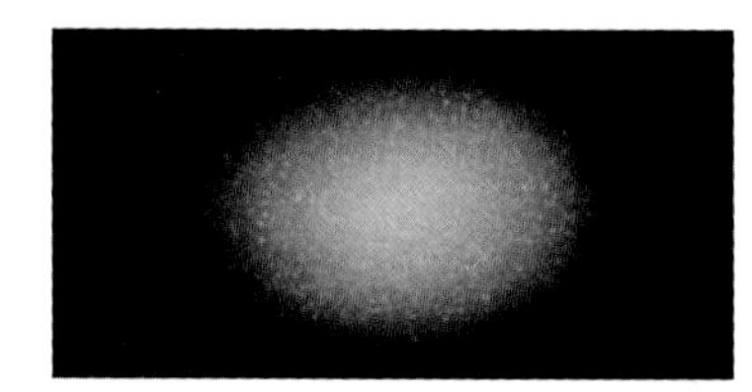

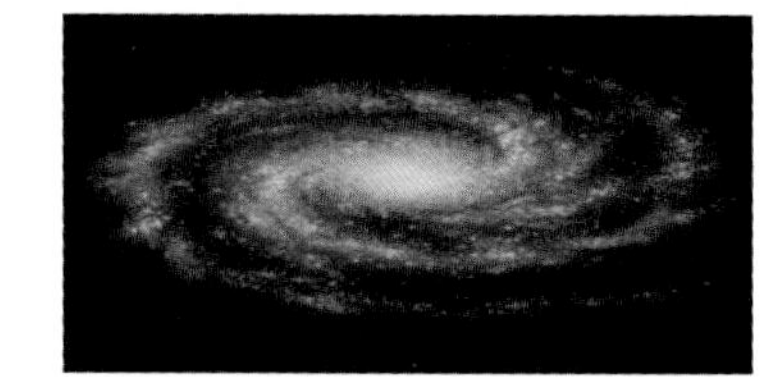

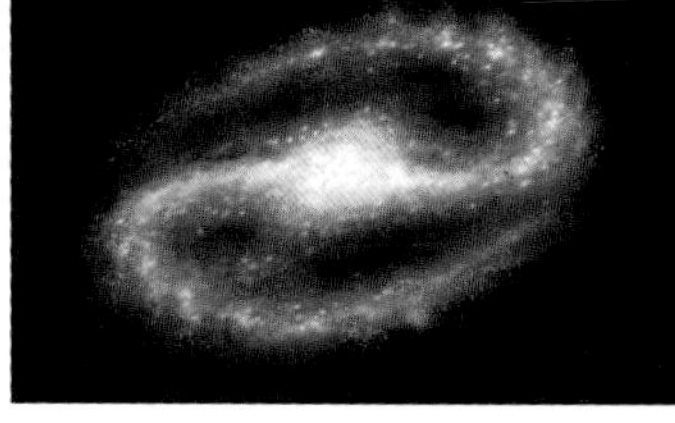

Types of galaxies
Galaxies exist in several forms, including elliptical (*top left*); spiral, with curved arms consisting of young stars (*middle*); and barred spiral, with a bar of stars through the center (*bottom*). Star-filled arms stretch from the bar.

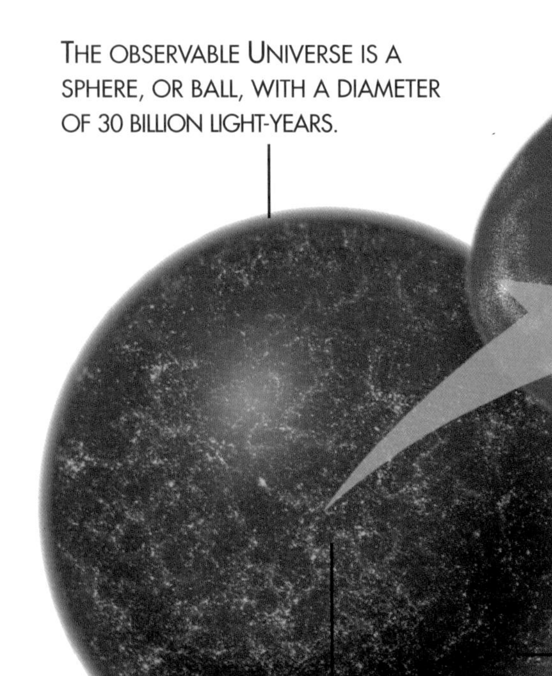

ANDROMEDA IS THE CLOSEST SPIRAL GALAXY TO THE MILKY WAY.
THE OBSERVABLE UNIVERSE IS A SPHERE, OR BALL, WITH A DIAMETER OF 30 BILLION LIGHT-YEARS.
OUR GALAXY, THE MILKY WAY, IS PART OF A SMALL CLUSTER OF ABOUT THIRTY GALAXIES KNOWN AS THE LOCAL GROUP. THE MILKY WAY IS SURROUNDED BY SMALLER GALAXIES.
SUPERCLUSTERS OF GALAXIES CAN BE BILLIONS OF LIGHT-YEARS LONG. THEY CONTAIN MANY CLUSTERS OF GALAXIES.
ONE GALAXY CLUSTER CAN CONTAIN THOUSANDS OF GALAXIES.
GIGANTIC FILAMENTS ARE MADE UP OF SUPERCLUSTERS.
MANY LARGE, EMPTY PLACES EXIST IN THE UNIVERSE TODAY.

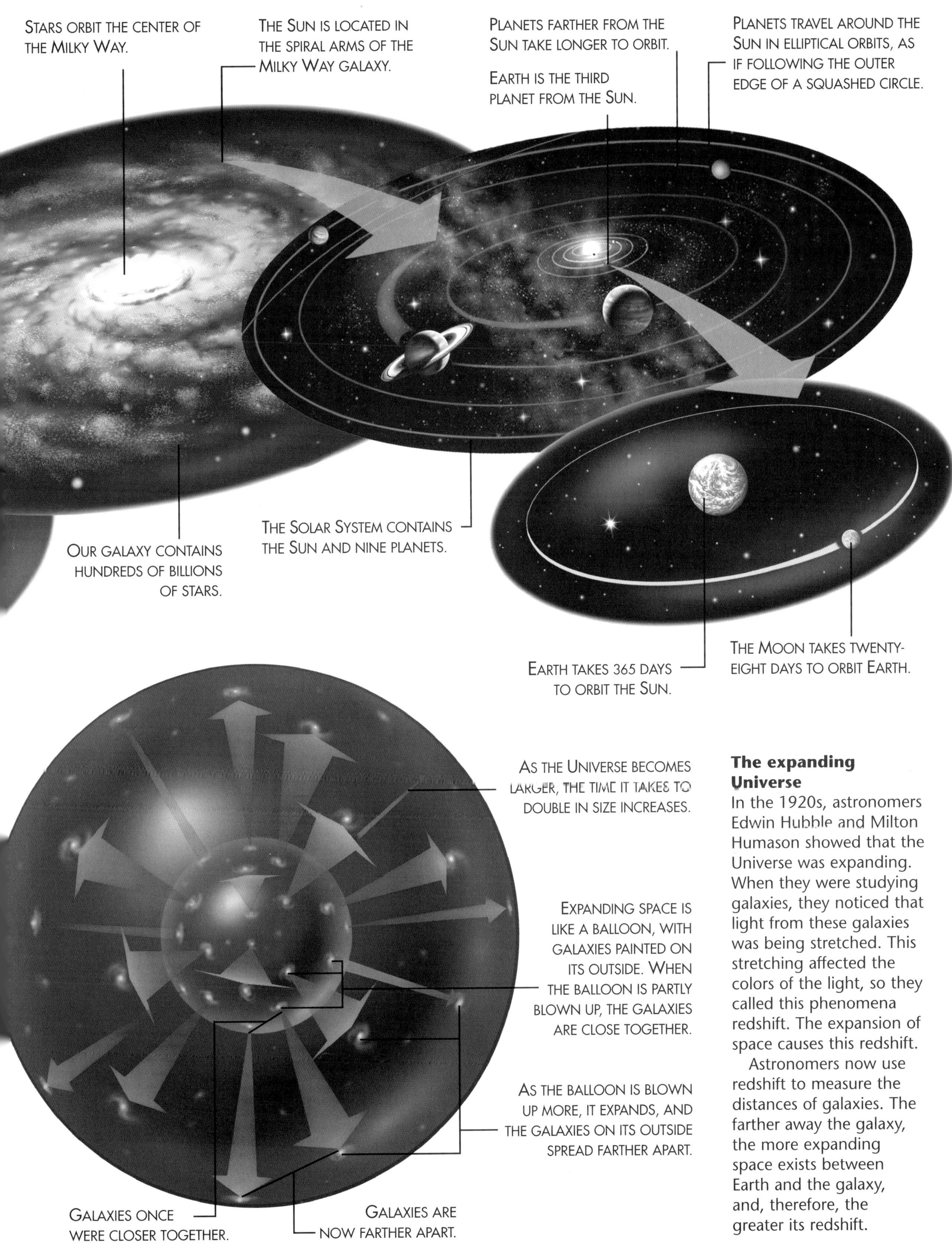
STARS ORBIT THE CENTER OF THE MILKY WAY.
THE SUN IS LOCATED IN THE SPIRAL ARMS OF THE MILKY WAY GALAXY.
PLANETS FARTHER FROM THE SUN TAKE LONGER TO ORBIT.
EARTH IS THE THIRD PLANET FROM THE SUN.
PLANETS TRAVEL AROUND THE SUN IN ELLIPTICAL ORBITS, AS IF FOLLOWING THE OUTER EDGE OF A SQUASHED CIRCLE.
OUR GALAXY CONTAINS HUNDREDS OF BILLIONS OF STARS.
THE SOLAR SYSTEM CONTAINS THE SUN AND NINE PLANETS.
EARTH TAKES 365 DAYS TO ORBIT THE SUN.
THE MOON TAKES TWENTY-EIGHT DAYS TO ORBIT EARTH.
AS THE UNIVERSE BECOMES LARGER, THE TIME IT TAKES TO DOUBLE IN SIZE INCREASES.
EXPANDING SPACE IS LIKE A BALLOON, WITH GALAXIES PAINTED ON ITS OUTSIDE. WHEN THE BALLOON IS PARTLY BLOWN UP, THE GALAXIES ARE CLOSE TOGETHER.
AS THE BALLOON IS BLOWN UP MORE, IT EXPANDS, AND THE GALAXIES ON ITS OUTSIDE SPREAD FARTHER APART.
GALAXIES ONCE WERE CLOSER TOGETHER.
GALAXIES ARE NOW FARTHER APART.

The expanding Universe

In the 1920s, astronomers Edwin Hubble and Milton Humason showed that the Universe was expanding. When they were studying galaxies, they noticed that light from these galaxies was being stretched. This stretching affected the colors of the light, so they called this phenomena redshift. The expansion of space causes this redshift.

Astronomers now use redshift to measure the distances of galaxies. The farther away the galaxy, the more expanding space exists between Earth and the galaxy, and, therefore, the greater its redshift.

The Solar System

A solar system consists of a star and all the objects that orbit it. Earth belongs to a family of nine planets that orbit the Sun, a star. The planets and the Sun in our Solar System formed about 4.5 billion years ago. Mercury, Venus, Earth, and Mars are the four inner planets. They are all made of rock, but only Earth has an atmosphere in which humans can breathe. An atmosphere is made up of the gases surrounding a planet. Beyond Mars is the asteroid belt, which contains millions of rocks. The four outer planets orbit the Sun on the other side of the asteroid belt. Jupiter, Saturn, Uranus, and Neptune are gas giants with very thick atmospheres and no solid surfaces. The smallest planet, Pluto, orbits beyond the gas giants. This rocky planet is always cold because it is so far from the Sun. Astronomers believe that more planets exist around other stars in our galaxy.

Orbits

The Sun's gravity keeps objects in the Solar System from floating off into space. The Sun contains so much matter that the pull of its gravity is very strong. This force causes the planets to move in orbits around it. These orbits are elliptical in shape.

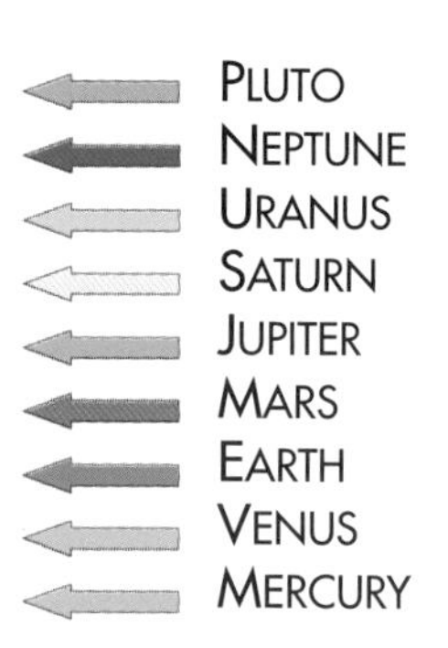

PLUTO
NEPTUNE
URANUS
SATURN
JUPITER
MARS
EARTH
VENUS
MERCURY

PLUTO IS THE FARTHEST PLANET FROM THE SUN. IT IS COVERED IN ICE.

SOME ASTRONOMERS BELIEVE A TENTH PLANET EXISTS BEYOND PLUTO'S ORBIT.

THE RINGS OF SATURN CONSIST OF TINY PEBBLES AND DUST.

URANUS IS TILTED ON ITS SIDE.

NEPTUNE'S ORBIT IS ELLIPTICAL.

JUPITER IS THE LARGEST PLANET IN THE SOLAR SYSTEM.

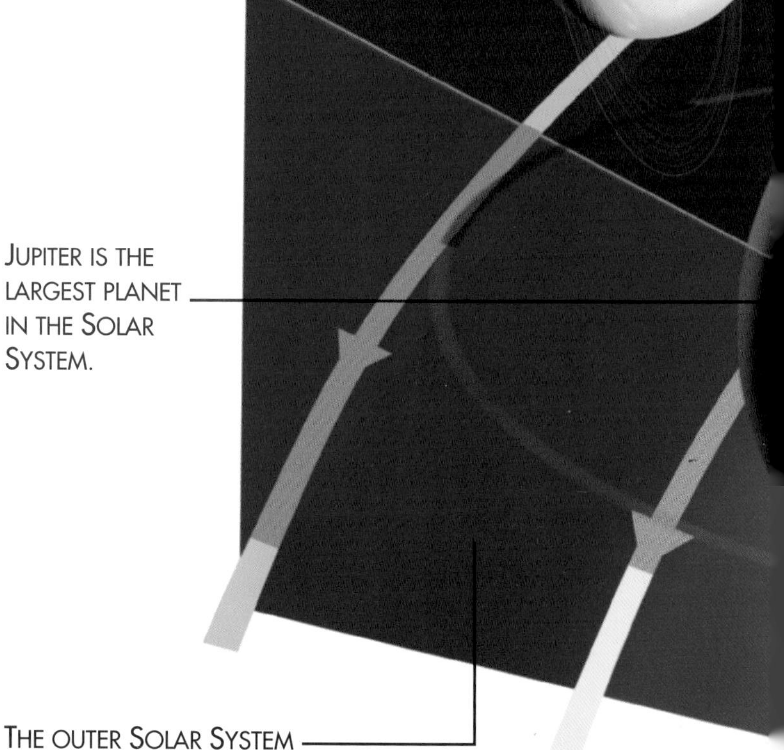

THE OUTER SOLAR SYSTEM

ASTEROIDS WERE DISCOVERED BY ASTRONOMERS WHO WERE LOOKING FOR A PLANET BETWEEN MARS AND JUPITER.

The Milky Way

The Sun is one of billions of stars that make up our galaxy, the Milky Way. The Milky Way has a flat, spiral shape. The Sun is in one of its spiral arms, closer to the galaxy's edge than to its center. Its location is shown by the red dots below. The Sun takes 226 million years to orbit the center of the galaxy. Our galaxy is one of billions of galaxies in the Universe.

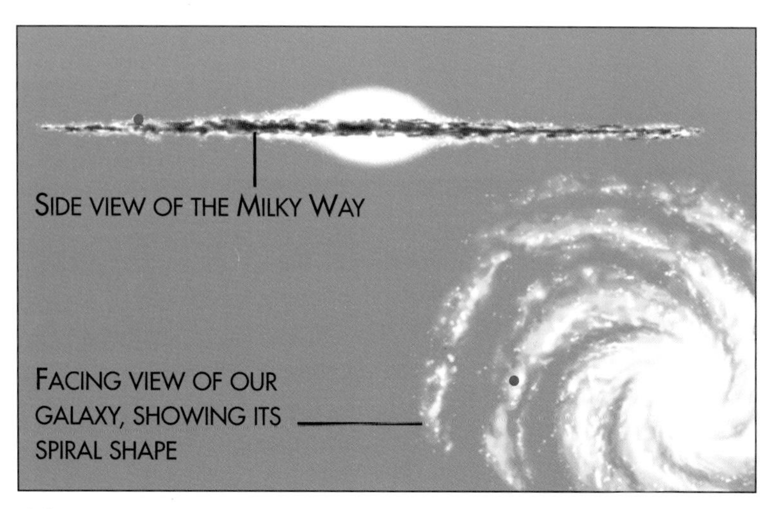

SIDE VIEW OF THE MILKY WAY
FACING VIEW OF OUR GALAXY, SHOWING ITS SPIRAL SHAPE

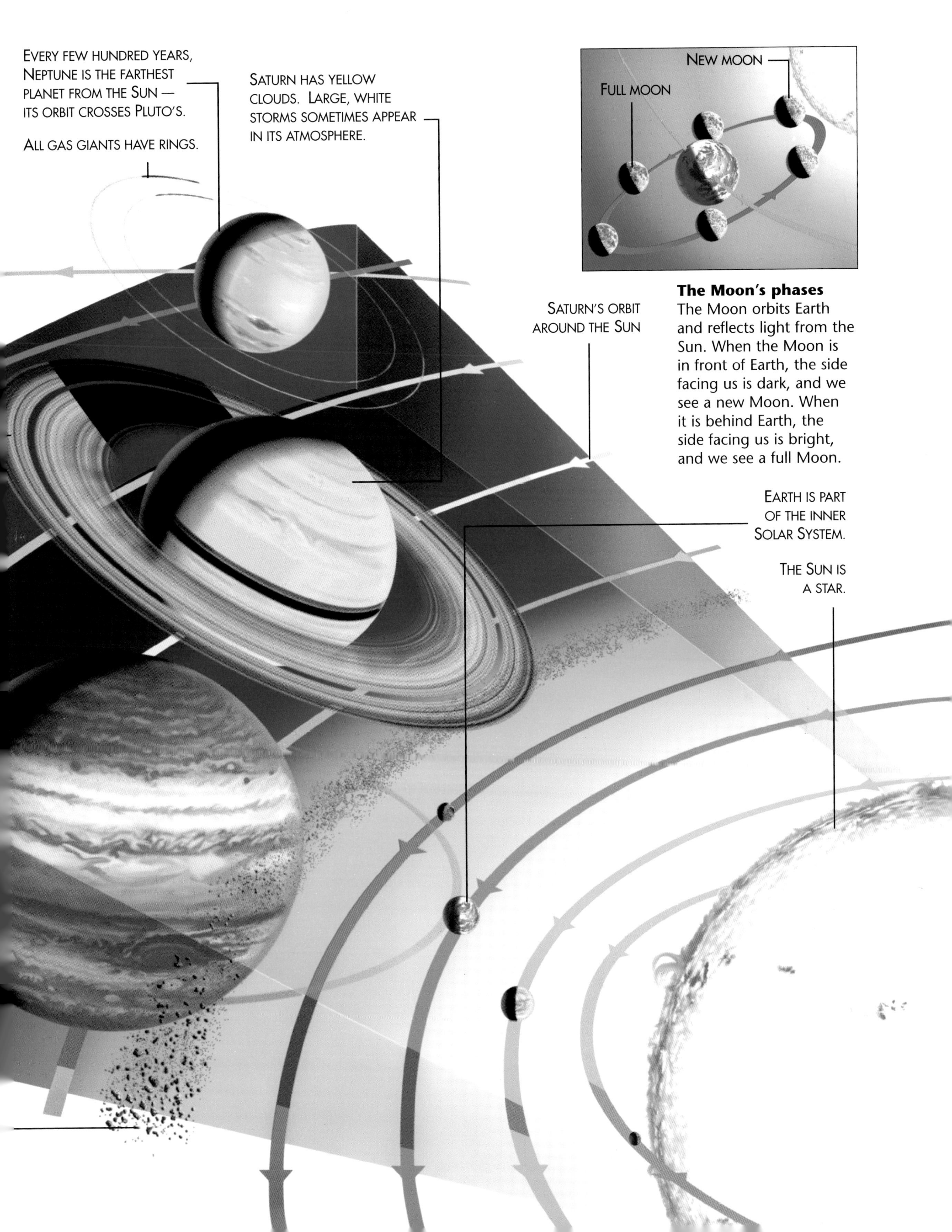
EVERY FEW HUNDRED YEARS, NEPTUNE IS THE FARTHEST PLANET FROM THE SUN — ITS ORBIT CROSSES PLUTO'S.
ALL GAS GIANTS HAVE RINGS.
SATURN HAS YELLOW CLOUDS. LARGE, WHITE STORMS SOMETIMES APPEAR IN ITS ATMOSPHERE.
NEW MOON
FULL MOON
SATURN'S ORBIT AROUND THE SUN
EARTH IS PART OF THE INNER SOLAR SYSTEM.
THE SUN IS A STAR.

The Moon's phases
The Moon orbits Earth and reflects light from the Sun. When the Moon is in front of Earth, the side facing us is dark, and we see a new Moon. When it is behind Earth, the side facing us is bright, and we see a full Moon.

The Planets

All nine planets in the Solar System orbit the Sun. Each one has a different surface structure, temperature, and atmosphere. Many of the planets have moons that travel around them. Jupiter, Saturn, Uranus, and Neptune also have rings.

Matter inside planets and moons exists in layers. The densest material is at the center. It is a metal, usually iron. The lighter, outer material consists of rocks. A planet's lightest material is gas. Gas forms an atmosphere above a planet's surface.

These layers show how planets and moons formed. Long ago, new planets hit each other. This made some of the rocks in the planets melt. While molten, their heavier material sank to their centers. Lighter material floated to their surfaces.

Moons
About sixty-seven moons exist in the Solar System. Earth's Moon is rocky with a small iron core (1). Most other moons orbit the gas giants. Jupiter's moons include Europa (2), which may be hiding an ocean beneath its icy crust, and Io (3), which has the most volcanoes in the Solar System. Clouds are always covering Titan (4), Saturn's largest moon.

A THIN, DUSTY RING CIRCLES JUPITER.

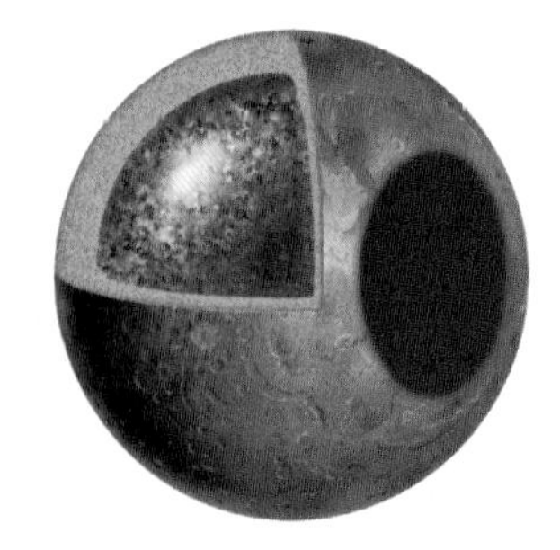

PLUTO'S MOON, CHARON, WAS DISCOVERED IN 1978.

SOMETIMES, CHARON BLOCKS THE SUN'S LIGHT, CASTING A SHADOW ON PLUTO.

Pluto
The tiniest planet is Pluto (*above*). It consists of ice surrounding a large core of rock and iron. A moon called Charon orbits it. Here, Charon is shown eclipsing the Sun.

Uranus
This gas giant (*lower right*) has a thick atmosphere of gas over an even thicker area of gas, ice, and rock particles. It may have a rocky core about the size of Earth at its center.

THE NARROW RINGS AROUND URANUS ARE MADE OF SMALL, DARK DUST PARTICLES.

Jupiter
The diameter of Jupiter (*above*) is eleven times greater than that of Earth. Below the surface of Jupiter, a mixture of hydrogen and helium gas is squeezed so tight that it acts like a liquid. Below this, the mixture is squeezed even more and behaves like a liquid metal. A rock and iron core five times larger than Earth may make up the planet's center.

Neptune
The planet Neptune (*left*) is about the same size as Uranus. It also has a similar inner structure. Neptune has patchy rings with thick and thin areas.

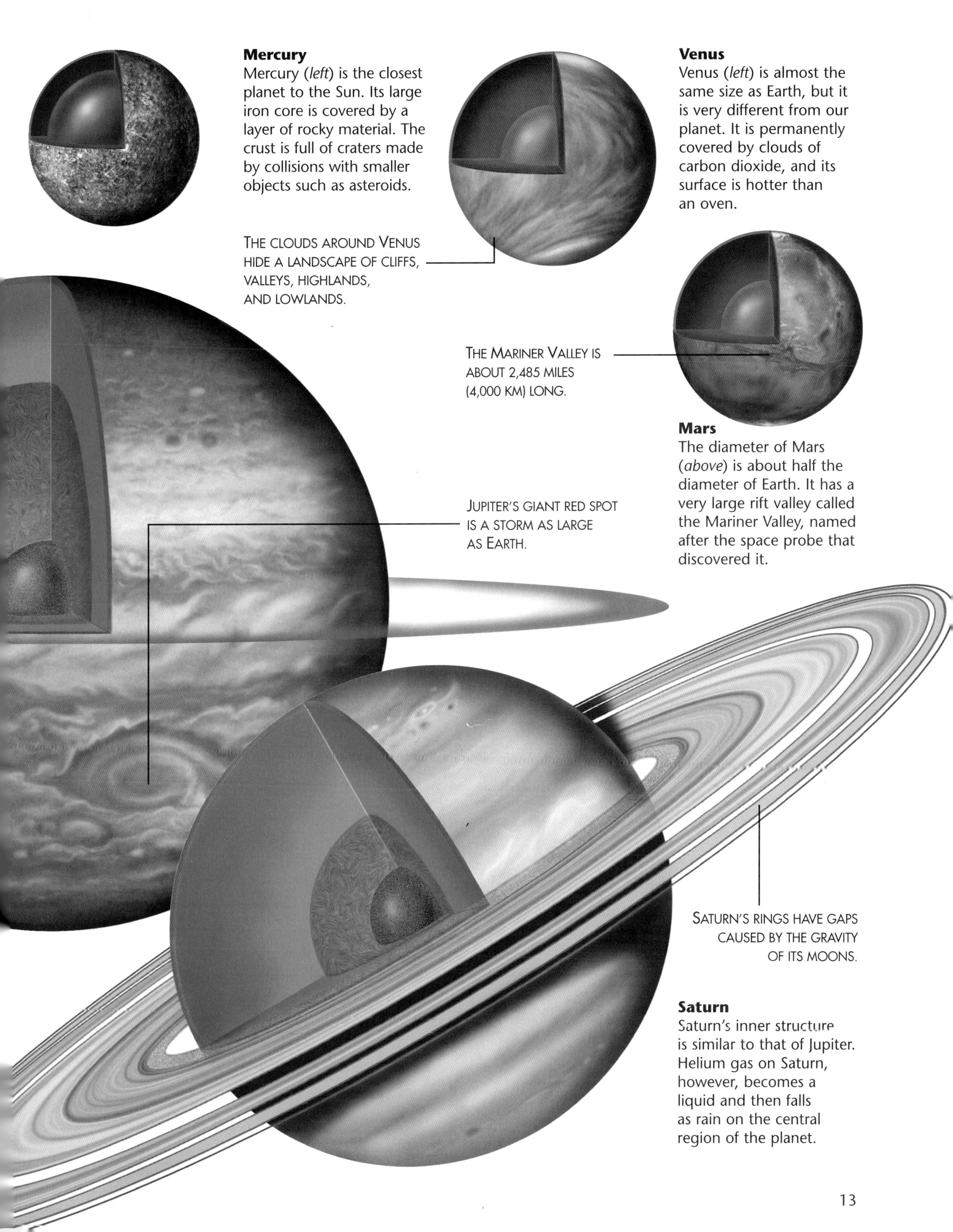
THE CLOUDS AROUND VENUS HIDE A LANDSCAPE OF CLIFFS, VALLEYS, HIGHLANDS, AND LOWLANDS.
THE MARINER VALLEY IS ABOUT 2,485 MILES (4,000 KM) LONG.
JUPITER'S GIANT RED SPOT IS A STORM AS LARGE AS EARTH.
SATURN'S RINGS HAVE GAPS CAUSED BY THE GRAVITY OF ITS MOONS.

Mercury

Mercury (*left*) is the closest planet to the Sun. Its large iron core is covered by a layer of rocky material. The crust is full of craters made by collisions with smaller objects such as asteroids.

Venus

Venus (*left*) is almost the same size as Earth, but it is very different from our planet. It is permanently covered by clouds of carbon dioxide, and its surface is hotter than an oven.

Mars

The diameter of Mars (*above*) is about half the diameter of Earth. It has a very large rift valley called the Mariner Valley, named after the space probe that discovered it.

Saturn

Saturn's inner structure is similar to that of Jupiter. Helium gas on Saturn, however, becomes a liquid and then falls as rain on the central region of the planet.

Planet Earth

We live on the planet Earth. About 4.5 billion years ago, Earth was just a sphere of lava, or molten rock. Heavy elements, such as iron, sank to the planet's center, and lighter elements floated to its outer edge. Scientists think comets crashing into Earth brought water and atmospheric gases to the planet. Comets are balls of ice and rock, often surrounded by gas.

Earth has several layers. Its center is a core of iron. A layer of rock, called the mantle, surrounds this core. Part of this layer is molten rock, or lava. Heat makes the molten rock flow in large circles. The rocky upper mantle and Earth's outer layer, or crust, are broken into pieces called plates, which float on the molten rock. Earthquakes occur and volcanoes erupt when these plates collide.

The Earth's core
The Earth's core contains one-third of the planet's mass, or weight. Recently, geologists listened to the way sounds travel through Earth and discovered that the core is divided into two regions. The outer core consists of liquid iron. The inner core is solid iron that is 13,046° F (7,230° C).

WHERE TWO OCEAN PLATES COLLIDE, ONE PLATE IS FORCED DOWNWARD.

HEAT FROM INSIDE EARTH CAUSES THE MANTLE'S MOLTEN ROCK TO MOVE IN CIRCULAR PATTERNS.

THE THIN CRUST IS MADE OF PLATES OF SOLIDIFIED ROCK. OCEAN PLATES FORM THE OCEAN FLOOR; CONTINENTAL PLATES FORM LAND.

CONVECTION IN THE OUTER CORE CAUSES EARTH TO GENERATE A MAGNETIC FIELD.

INNER MANTLE ROCKS ARE PRESSED SOLID BY THE WEIGHT OF MATERIAL ABOVE THEM.

ROCKS ARE MOLTEN IN THE OUTER AREA OF THE MANTLE.

MOUNTAINS ARE OFTEN BUILT BY COLLIDING PLATES.

WATER VAPOR IN EARTH'S ATMOSPHERE FORMS CLOUDS.

THE SOUTH POLE IS OPPOSITE THE NORTH POLE.

ON ITS AXIS, EARTH ROTATES TO PRODUCE NIGHT AND DAY.

The seasons
Earth is tilted at an angle as it orbits the Sun. When the North Pole is tilted toward the Sun, the Northern Hemisphere has summer (1). Six months later, the North Pole is tilted away from the Sun, and it is winter there (2). Spring (3) and autumn (4) occur in between.

WHEN IT IS WINTER IN THE NORTH, IT IS SUMMER IN THE SOUTH.

1
2
3
4

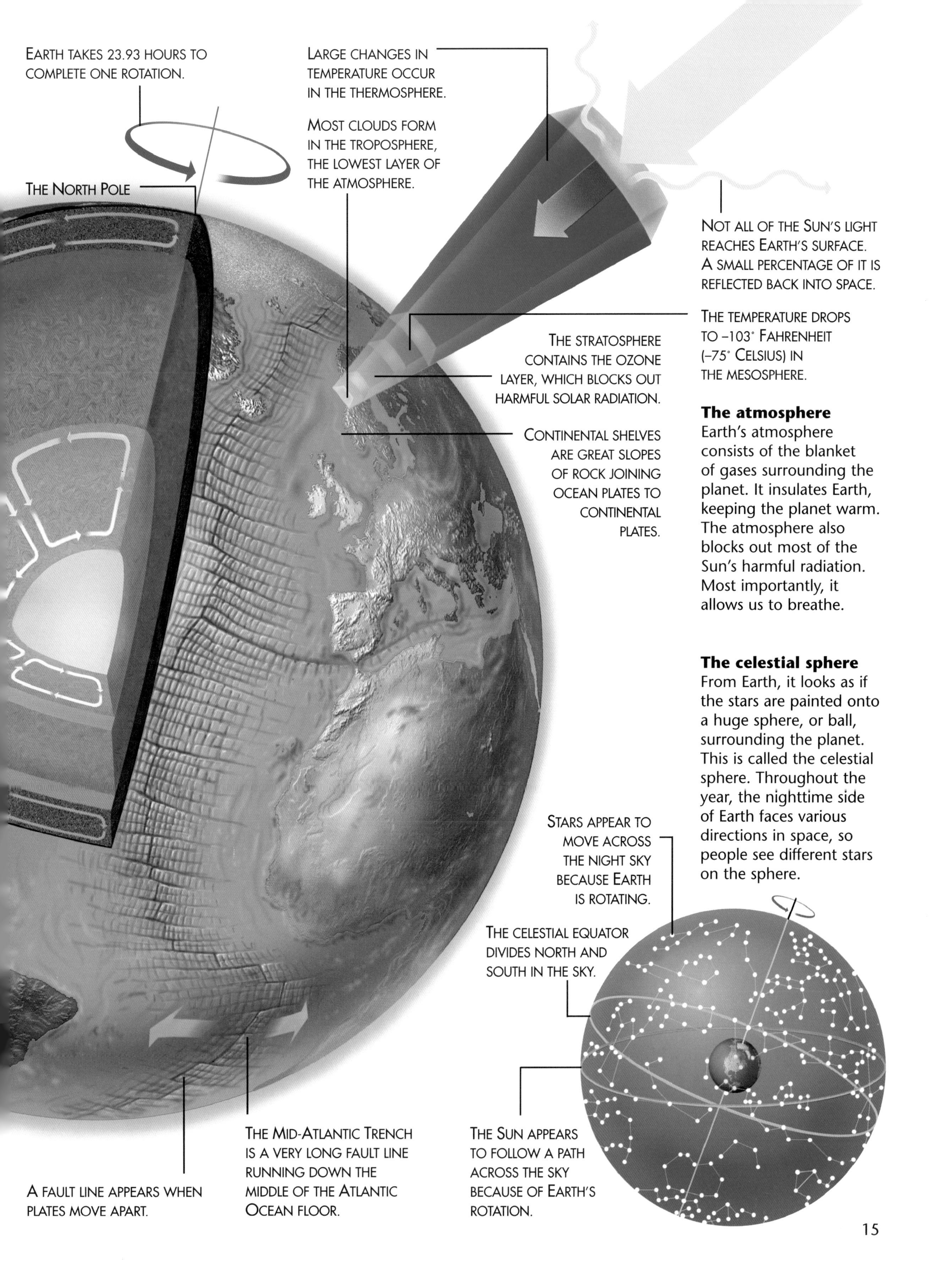
EARTH TAKES 23.93 HOURS TO COMPLETE ONE ROTATION.
THE NORTH POLE
LARGE CHANGES IN TEMPERATURE OCCUR IN THE THERMOSPHERE.
MOST CLOUDS FORM IN THE TROPOSPHERE, THE LOWEST LAYER OF THE ATMOSPHERE.
NOT ALL OF THE SUN'S LIGHT REACHES EARTH'S SURFACE. A SMALL PERCENTAGE OF IT IS REFLECTED BACK INTO SPACE.
THE TEMPERATURE DROPS TO –103° FAHRENHEIT (–75° CELSIUS) IN THE MESOSPHERE.
THE STRATOSPHERE CONTAINS THE OZONE LAYER, WHICH BLOCKS OUT HARMFUL SOLAR RADIATION.
CONTINENTAL SHELVES ARE GREAT SLOPES OF ROCK JOINING OCEAN PLATES TO CONTINENTAL PLATES.
STARS APPEAR TO MOVE ACROSS THE NIGHT SKY BECAUSE EARTH IS ROTATING.
THE CELESTIAL EQUATOR DIVIDES NORTH AND SOUTH IN THE SKY.
A FAULT LINE APPEARS WHEN PLATES MOVE APART.
THE MID-ATLANTIC TRENCH IS A VERY LONG FAULT LINE RUNNING DOWN THE MIDDLE OF THE ATLANTIC OCEAN FLOOR.
THE SUN APPEARS TO FOLLOW A PATH ACROSS THE SKY BECAUSE OF EARTH'S ROTATION.

The atmosphere
Earth's atmosphere consists of the blanket of gases surrounding the planet. It insulates Earth, keeping the planet warm. The atmosphere also blocks out most of the Sun's harmful radiation. Most importantly, it allows us to breathe.

The celestial sphere
From Earth, it looks as if the stars are painted onto a huge sphere, or ball, surrounding the planet. This is called the celestial sphere. Throughout the year, the nighttime side of Earth faces various directions in space, so people see different stars on the sphere.

Rocket Power

The concept of rockets can be traced to Ancient Greeks in the fourth century B.C. People did not learn how to build them, however, until the thirteenth century A.D., when the Chinese built rockets with gunpowder and bamboo tubes. The Chinese attached these rockets to arrows and fired them. In times of peace, rockets were used as fireworks. During war, they were used as weapons.

The principle behind rockets is that every action creates an equal and opposite reaction. This is one of Isaac Newton's laws of motion. In rockets, the action is an explosion, which produces gas that is directed out of a tube. The gas leaving the tube causes a reaction — it pushes the tube in the opposite direction.

Rockets blast off from launchpads, taking objects such as satellites into space. In the future, spaceplanes will use rocket engines to take off like airplanes from runways. They will fly so high that they will reach space.

SATELLITES TAKEN INTO ORBIT ARE KNOWN AS THE PAYLOAD.

THIS SATELLITE IS A TELESCOPE THAT WILL BE PLACED IN A LOW EARTH ORBIT.

PAYLOADS MUST BE DESIGNED TO FIT INTO THE ROCKET FROM WHICH THEY WILL BE LAUNCHED.

SATELLITES MAY HAVE SMALL ROCKETS OF THEIR OWN TO BLAST THEM INTO THEIR FINAL ORBITS.

ARIANE 4 IS A EUROPEAN-BUILT ROCKET, LAUNCHED FROM SOUTH AMERICA. IT IS 197 FEET (60 M) LONG.

Stage rockets

Although early rockets contained single engines, scientists soon realized that rockets should be built in stages. A *stage* is a part of a rocket that can fall away when it is no longer needed. This reduces the mass of a rocket, so the remaining rocket engines can accelerate it more easily.

Three-stage rockets have the most effective design. The rocket's first stage lifts it from the ground. When its fuel tanks are empty, the first stage falls off, and the second stage takes over. Finally, a third stage sends the rocket into orbit.

History of rockets

Early in the twentieth century, an American named Robert Goddard launched the first modern rocket (1). It flew to a height of 41 feet (12.5 m).

Rockets have always been used as weapons. In the 1940s, the Germans developed the *V-2* rocket, which was launched over London (2). Many other countries have also developed rockets. Russia built D-class vehicles to launch their *Soyuz* manned space missions (3). The *Saturn V* rocket (4) was developed by the United States for the *Apollo* Moon missions. The present-day Space Shuttle (5) is the first reusable spacecraft.

THE LARGEST ROCKET EVER BUILT WAS SATURN V, WHICH WENT TO THE MOON. IT WAS MORE THAN 360 FEET (110 METERS) LONG.

THE SPACE SHUTTLE DOES NOT GO AS FAR AS THE MOON, SO IT IS SMALLER THAN SATURN V.

1

2

3

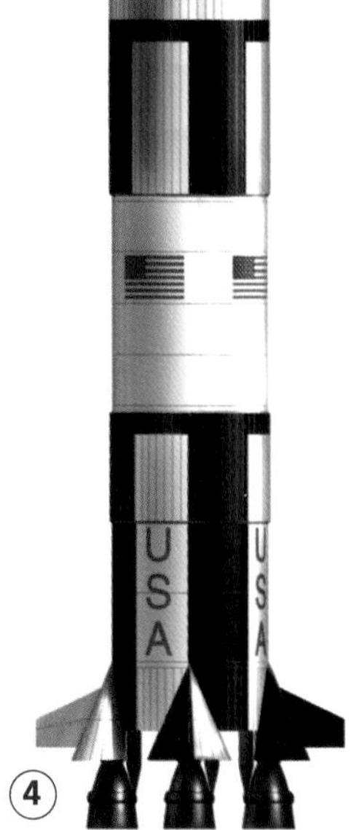
4
USA

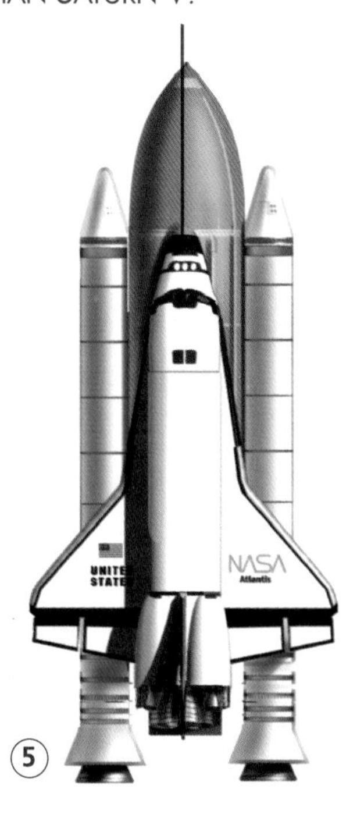
5
NASA

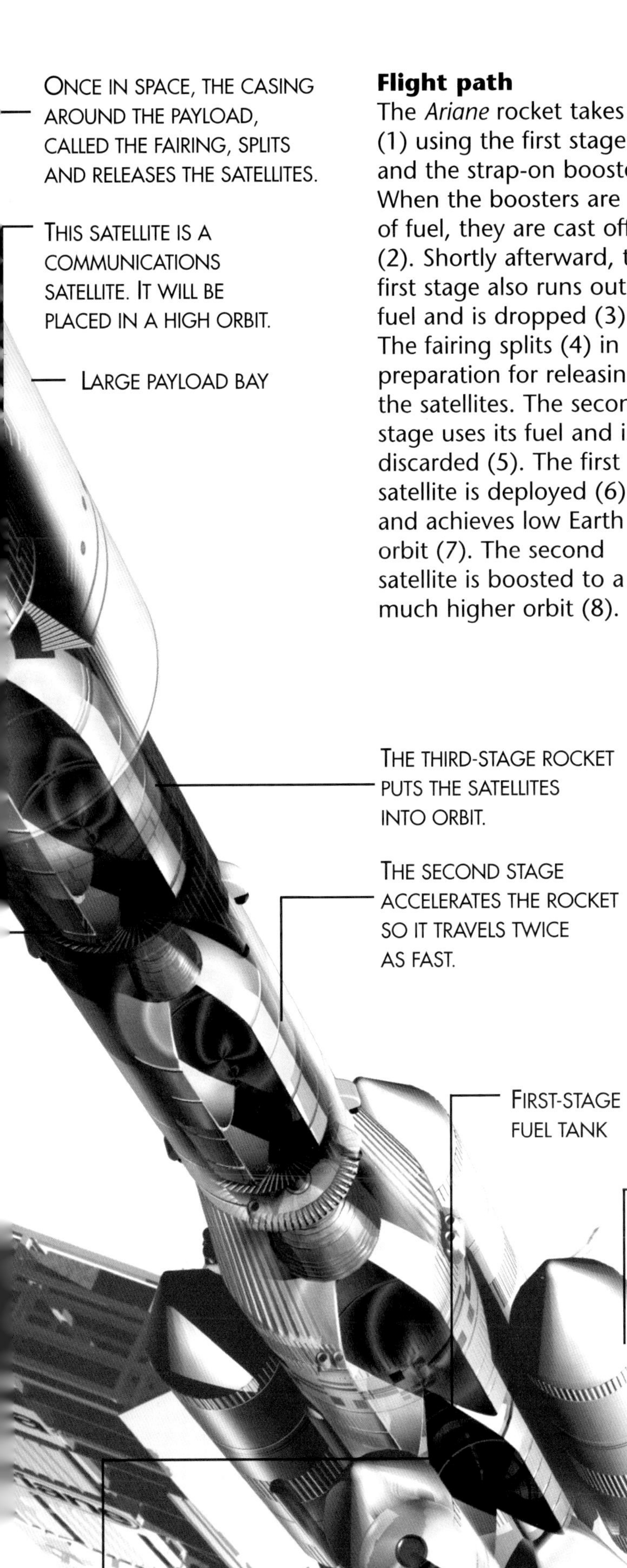

ONCE IN SPACE, THE CASING AROUND THE PAYLOAD, CALLED THE FAIRING, SPLITS AND RELEASES THE SATELLITES.
THIS SATELLITE IS A COMMUNICATIONS SATELLITE. IT WILL BE PLACED IN A HIGH ORBIT.
LARGE PAYLOAD BAY
THE THIRD-STAGE ROCKET PUTS THE SATELLITES INTO ORBIT.
THE SECOND STAGE ACCELERATES THE ROCKET SO IT TRAVELS TWICE AS FAST.
FIRST-STAGE FUEL TANK
THE FIRST STAGE IS HELPED BY STRAP-ON BOOSTER ROCKETS.
FIRST-STAGE OXYGEN TANK
FIRST-STAGE ROCKET ENGINES

Flight path
The *Ariane* rocket takes off (1) using the first stage and the strap-on boosters. When the boosters are out of fuel, they are cast off (2). Shortly afterward, the first stage also runs out of fuel and is dropped (3). The fairing splits (4) in preparation for releasing the satellites. The second stage uses its fuel and is discarded (5). The first satellite is deployed (6) and achieves low Earth orbit (7). The second satellite is boosted to a much higher orbit (8).

TO ACHIEVE LIFTOFF, A ROCKET MUST PRODUCE ENOUGH THRUST TO OVERCOME THE FORCE OF GRAVITY.
LIQUID HYDROGEN FUEL IS SUPPLIED TO THE ROCKET FROM THE FUEL TANK.
A SUPPLY OF OXYGEN HELPS THE FUEL IGNITE.
THE COMBUSTION CHAMBER LIGHTS THE FUEL AND DIRECTS THE THRUST BEHIND THE ROCKET. THIS CAUSES THE ROCKET TO MOVE UPWARD.

Rocket fuel
Rocket engines ignite fuel to produce thrust. Liquid oxygen mixed with another liquid fuel, such as hydrogen or fluorine, produces more thrust than a single fuel could. In the future, rockets may use three fuels to gain even more energy.

Early Missions

The former Soviet Union was the first country to successfully send a person into space. Astronaut Yuri Gagarin flew in the rocket *Vostok 1* on April 12, 1961. His flight lasted only 108 minutes. During that time, he completed one orbit around Earth and then returned safely. The rocket that launched *Vostok 1* was originally a missile for carrying explosives.

The Soviets launched *Vostok 2* on August 6, 1961. This time, the astronaut stayed in space for a whole day.

A year later, the Soviets launched astronauts in *Vostok 3* and *4* with just twenty-four hours between the launches. Both spacecraft successfully returned to Earth a few days later. In 1963, *Vostok 5* and *6* were launched. *Vostok 6* carried Valentina Tereshkova, the first female astronaut to travel into space.

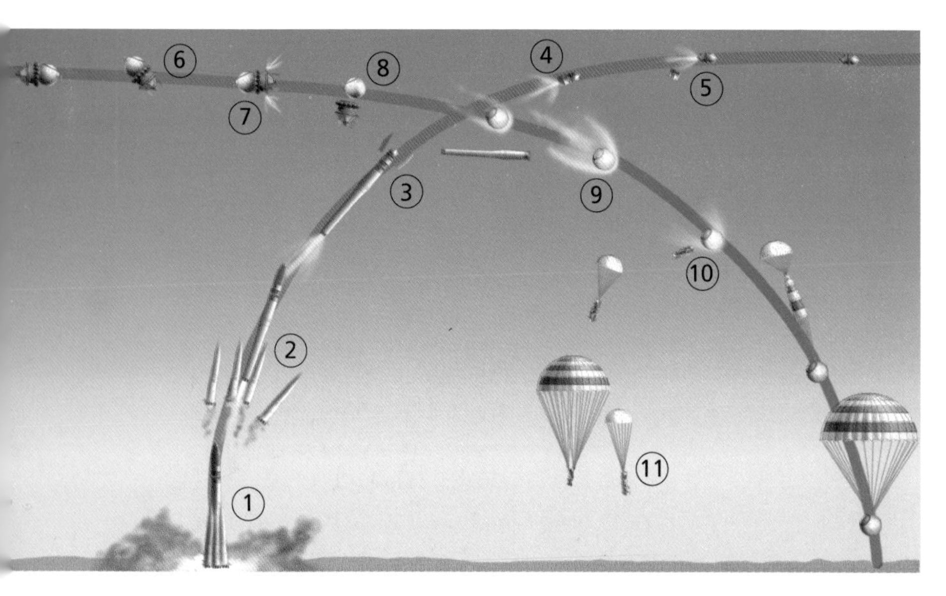

6
8
4
7
5
3
9
10
2
11
1

Flight profile
Vostok 1 blasts off (1) and drops the first-stage boosters (2). The nose casing is ejected as the second stage increases *Vostok's* speed (3). The final stage puts *Vostok 1* into orbit (4) and then separates (5). After orbiting once, it turns (6) and fires the retro-rocket (7). The instrument module is discarded (8), and *Vostok* reenters the atmosphere (9). The astronaut is ejected (10). Parachutes return the module and astronaut safely to Earth (11).

The instrument module
Scientists on the ground controlled the instrument module. It had rocket engines that could control the movements of *Vostok* once it was in orbit. Communication antennae were also attached to it. The module was ejected and allowed to burn up during reentry to Earth.

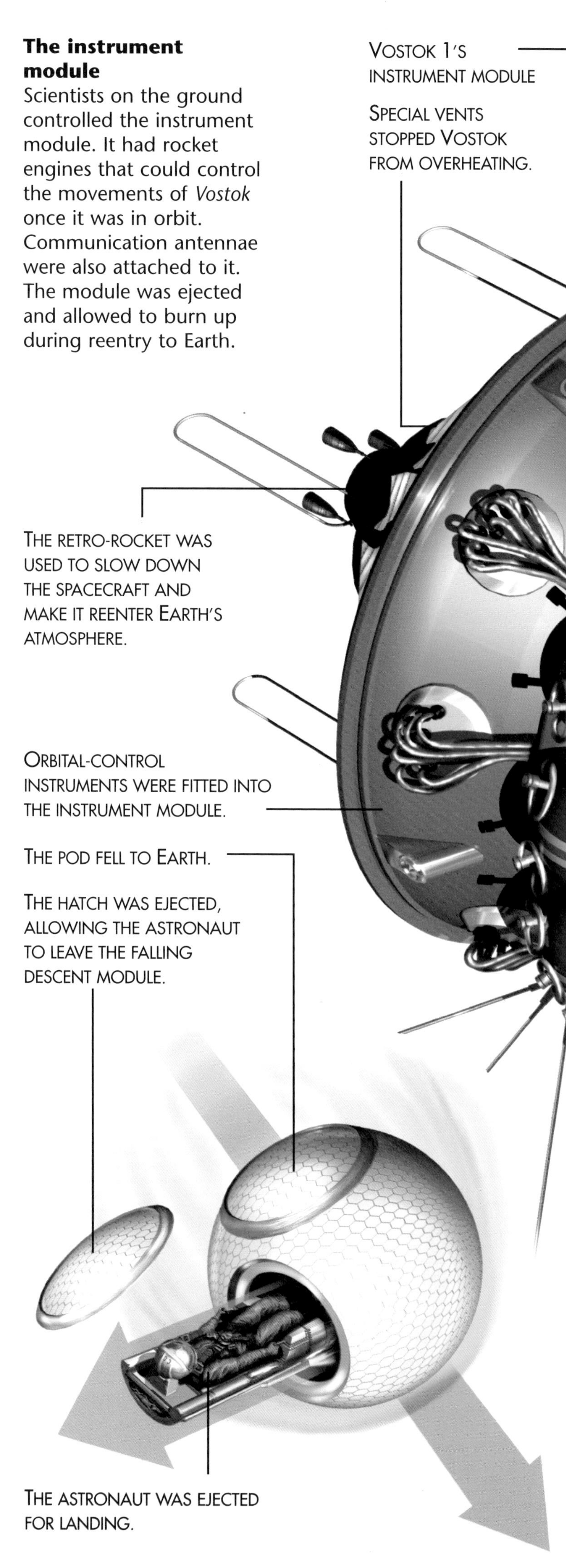

VOSTOK 1'S INSTRUMENT MODULE
SPECIAL VENTS STOPPED VOSTOK FROM OVERHEATING.
THE RETRO-ROCKET WAS USED TO SLOW DOWN THE SPACECRAFT AND MAKE IT REENTER EARTH'S ATMOSPHERE.
ORBITAL-CONTROL INSTRUMENTS WERE FITTED INTO THE INSTRUMENT MODULE.
THE POD FELL TO EARTH.
THE HATCH WAS EJECTED, ALLOWING THE ASTRONAUT TO LEAVE THE FALLING DESCENT MODULE.
THE ASTRONAUT WAS EJECTED FOR LANDING.

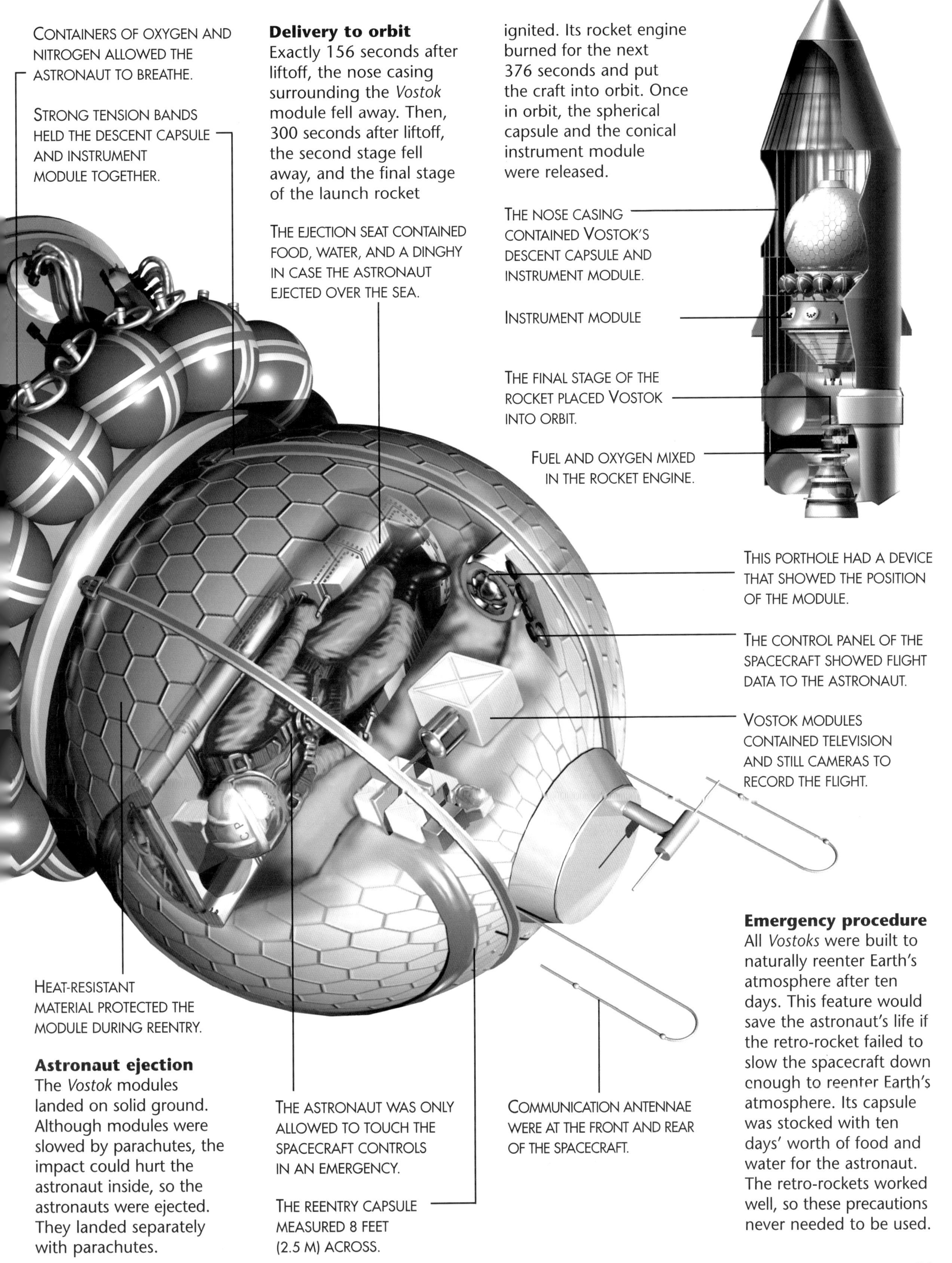
CONTAINERS OF OXYGEN AND NITROGEN ALLOWED THE ASTRONAUT TO BREATHE.
STRONG TENSION BANDS HELD THE DESCENT CAPSULE AND INSTRUMENT MODULE TOGETHER.
THE EJECTION SEAT CONTAINED FOOD, WATER, AND A DINGHY IN CASE THE ASTRONAUT EJECTED OVER THE SEA.
THE NOSE CASING CONTAINED VOSTOK'S DESCENT CAPSULE AND INSTRUMENT MODULE.
INSTRUMENT MODULE
THE FINAL STAGE OF THE ROCKET PLACED VOSTOK INTO ORBIT.
FUEL AND OXYGEN MIXED IN THE ROCKET ENGINE.
THIS PORTHOLE HAD A DEVICE THAT SHOWED THE POSITION OF THE MODULE.
THE CONTROL PANEL OF THE SPACECRAFT SHOWED FLIGHT DATA TO THE ASTRONAUT.
VOSTOK MODULES CONTAINED TELEVISION AND STILL CAMERAS TO RECORD THE FLIGHT.
HEAT-RESISTANT MATERIAL PROTECTED THE MODULE DURING REENTRY.
THE ASTRONAUT WAS ONLY ALLOWED TO TOUCH THE SPACECRAFT CONTROLS IN AN EMERGENCY.
THE REENTRY CAPSULE MEASURED 8 FEET (2.5 M) ACROSS.
COMMUNICATION ANTENNAE WERE AT THE FRONT AND REAR OF THE SPACECRAFT.

Delivery to orbit
Exactly 156 seconds after liftoff, the nose casing surrounding the *Vostok* module fell away. Then, 300 seconds after liftoff, the second stage fell away, and the final stage of the launch rocket ignited. Its rocket engine burned for the next 376 seconds and put the craft into orbit. Once in orbit, the spherical capsule and the conical instrument module were released.

Astronaut ejection
The *Vostok* modules landed on solid ground. Although modules were slowed by parachutes, the impact could hurt the astronaut inside, so the astronauts were ejected. They landed separately with parachutes.

Emergency procedure
All *Vostoks* were built to naturally reenter Earth's atmosphere after ten days. This feature would save the astronaut's life if the retro-rocket failed to slow the spacecraft down enough to reenter Earth's atmosphere. Its capsule was stocked with ten days' worth of food and water for the astronaut. The retro-rockets worked well, so these precautions never needed to be used.

Apollo Missions

Some of the greatest achievements of humankind are the space missions of the late 1960s and early 1970s. These missions placed humans on the Moon. On July 20, 1969, *Apollo 11* touched down on the surface of the Moon. Shortly afterward, astronaut Neil Armstrong became the first human being to walk on another world in the Solar System.

Seven manned missions were made to the Moon. Each mission involved three astronauts. In the six successful missions, one astronaut stayed in the command and service module, while two descended to the Moon's surface in the landing module.

Apollo 13 almost became a disaster when an electrical system short-circuited and caused an explosion, which destroyed an oxygen tank. The astronauts gave up landing on the Moon and returned safely to Earth. They used oxygen from the landing module to keep themselves alive.

Arriving at the Moon
When the *Apollo* spacecraft arrived at the Moon, the landing module and the command and service module were joined. After a retro-rocket was fired, both spacecraft entered an orbit around the Moon. Two astronauts went into the lander and fired another retro-rocket to separate the two modules. The landing module then drifted slowly down to the surface of the Moon.

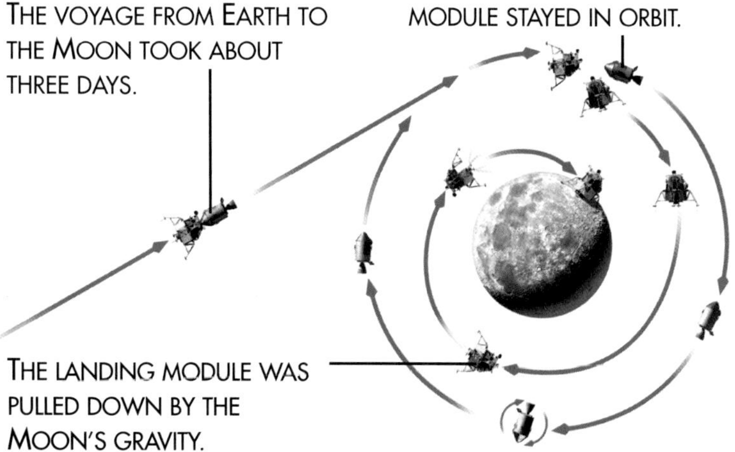
THE VOYAGE FROM EARTH TO THE MOON TOOK ABOUT THREE DAYS.
THE COMMAND AND SERVICE MODULE STAYED IN ORBIT.
THE LANDING MODULE WAS PULLED DOWN BY THE MOON'S GRAVITY.

UNITED STATES
THIS ENGINE WAS USED FOR ORBIT CHANGES.

Command and service module
The command and service module (*above*) was the astronauts' home during the voyages to and from the Moon. The conical section at the front is the command module, where the astronauts sat. It separated from the service module and reentered Earth's atmosphere when the mission was complete.

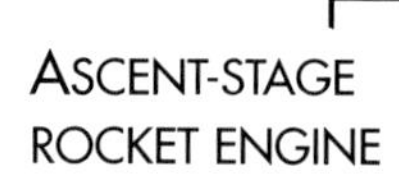
THIS RADAR DISH HELPED THE MODULES RECONNECT.
THE CREW USED A HATCH TO ENTER AND EXIT THE LANDING MODULE.
ASCENT-STAGE ROCKET ENGINE
OXYGEN FROM THIS TANK WAS MIXED WITH ROCKET FUEL AND FED TO THE MAIN DESCENT ENGINE.
THE CREW USED THIS LADDER TO CLIMB DOWN ONTO THE SURFACE OF THE MOON.
LANDING STRUTS ABSORBED THE SHOCK OF LANDING.
AS SOON AS THESE SENSORS TOUCHED THE MOON'S SURFACE, THE ENGINES WERE TURNED OFF.

SMALL ROCKETS, OR THRUSTERS, WERE USED TO CHANGE DIRECTION.

A DOCKING CLAMP GRABBED THE LANDER WHEN IT RETURNED FROM THE MOON.

THE RETURNING COMMAND MODULE SPLASHED DOWN IN THE OCEAN.

ANTENNAE ALLOWED THE MODULES TO COMMUNICATE.

THE DOCKING HATCH WAS AT THE TOP OF THE SPACECRAFT.

Back to Earth
When the astronauts finished their tasks on the Moon, the ascent stage of the landing module lifted off and rejoined the command and service modules. The astronauts then traveled back to Earth. The lander was left behind to fall back onto the Moon.

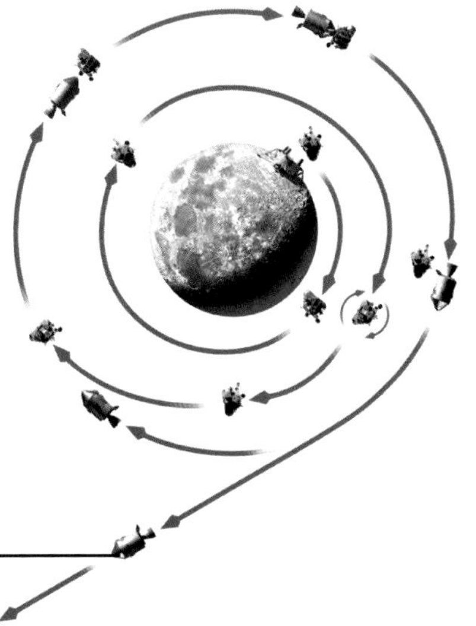

FIRING A ROCKET PUT THE COMMAND AND SERVICE MODULES IN AN ORBIT BACK TO EARTH.

The landing module
The landing module (*left*) was the small spacecraft that took astronauts to the surface of the Moon. It was split into two parts. Astronauts sat in the top part. Underneath this was a section used for equipment storage.

When it was time to leave the Moon, only the top part of the lander took off. This was called the ascent stage. Because the force of gravity is weaker on the Moon than on Earth, only a small rocket was needed to lift off.

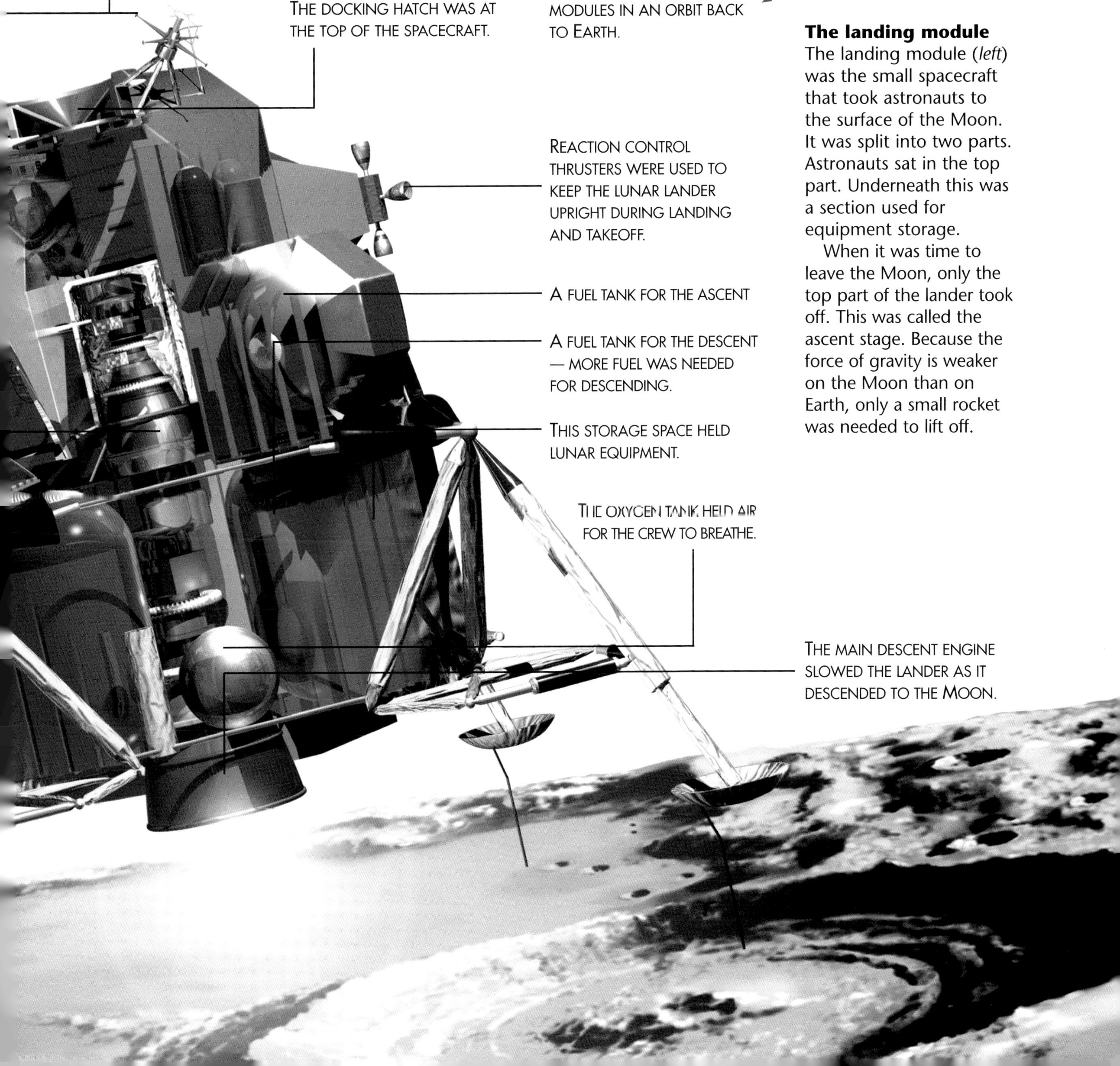

REACTION CONTROL THRUSTERS WERE USED TO KEEP THE LUNAR LANDER UPRIGHT DURING LANDING AND TAKEOFF.

A FUEL TANK FOR THE ASCENT

A FUEL TANK FOR THE DESCENT — MORE FUEL WAS NEEDED FOR DESCENDING.

THIS STORAGE SPACE HELD LUNAR EQUIPMENT.

THE OXYGEN TANK HELD AIR FOR THE CREW TO BREATHE.

THE MAIN DESCENT ENGINE SLOWED THE LANDER AS IT DESCENDED TO THE MOON.

On the Moon

Astronauts from *Apollo 11* were the first humans to walk on the Moon. They could only make one trip outside their lander. They spent 2.5 hours walking on the Moon before they climbed back into the lander.

On each of the following *Apollo* missions, astronauts spent more and more time on the Moon's surface. During the *Apollo 16* and *17* missions, they made several trips out of the lander and spent over twenty hours at work on the Moon's surface. *Apollo* missions stopped after the *Apollo 17* mission in December 1972. From 1998 to 1999, a space probe called the *Lunar Prospector* orbited the Moon to collect geological data. Scientists programmed it to crash on the Moon's surface, hoping it would break through to water. But none was found.

Moon walking

While they were on the surface of the Moon, the astronauts performed many tasks. They collected rock samples and also left equipment and experiments behind. These items sent information back to Earth long after the astronauts were gone.

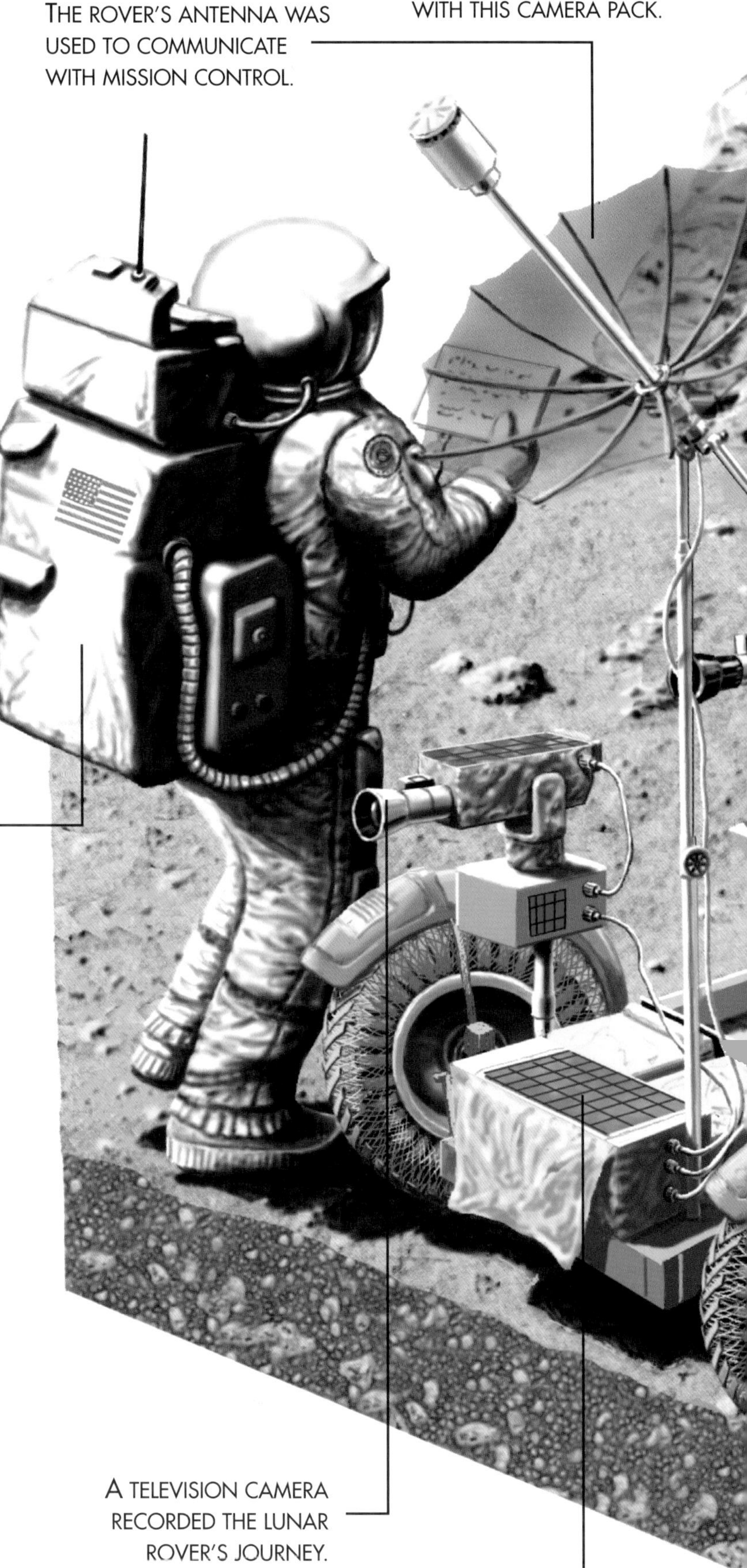
STILL PHOTOS WERE TAKEN WITH THIS CAMERA PACK.
THE ROVER'S ANTENNA WAS USED TO COMMUNICATE WITH MISSION CONTROL.
THE PLSS IS STRAPPED ONTO THE SPACE SUIT.
A TELEVISION CAMERA RECORDED THE LUNAR ROVER'S JOURNEY.
THE COMMUNICATION SYSTEM'S ELECTRONICS

Portable life support system

Because the Moon is so small, no air exists on its surface. This means that astronauts have to wear space suits to stay alive. The space suit was connected to a backpack called the Portable Life Support System (PLSS). The PLSS pumped oxygen (1) and water (2) into the space suits so the astronauts could breathe and keep their body temperature normal. The astronauts talked to each other with a radio (3). An emergency oxygen supply (4) was also included.

Surveyor probes
Before humans set foot on the Moon, *Surveyor* probes were sent there. Cameras on the probes took photographs of the surface. These pictures helped scientists choose *Apollo* landing sites.

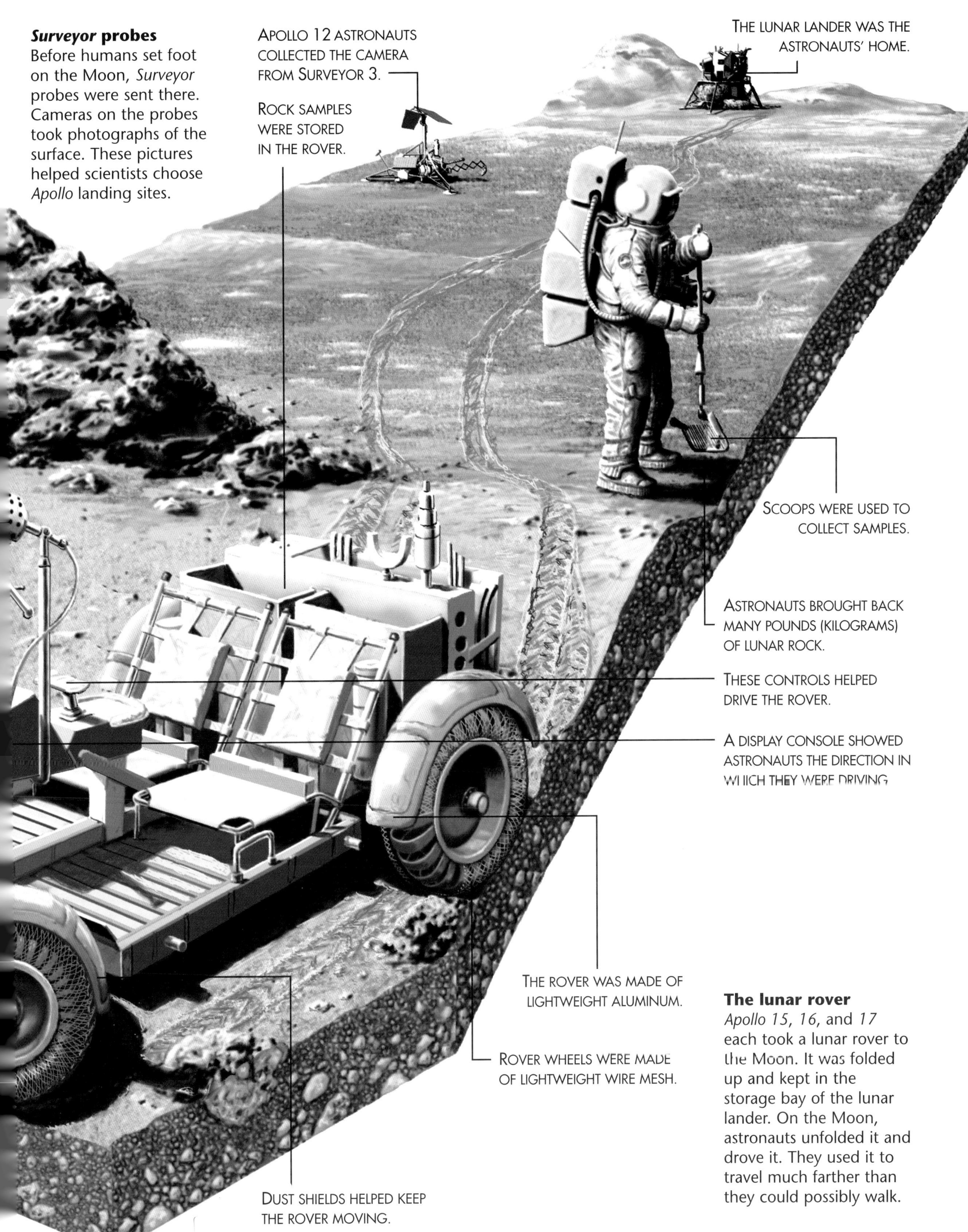

APOLLO 12 ASTRONAUTS COLLECTED THE CAMERA FROM SURVEYOR 3.
ROCK SAMPLES WERE STORED IN THE ROVER.
THE LUNAR LANDER WAS THE ASTRONAUTS' HOME.
SCOOPS WERE USED TO COLLECT SAMPLES.
ASTRONAUTS BROUGHT BACK MANY POUNDS (KILOGRAMS) OF LUNAR ROCK.
THESE CONTROLS HELPED DRIVE THE ROVER.
A DISPLAY CONSOLE SHOWED ASTRONAUTS THE DIRECTION IN WHICH THEY WERE DRIVING.
THE ROVER WAS MADE OF LIGHTWEIGHT ALUMINUM.
ROVER WHEELS WERE MADE OF LIGHTWEIGHT WIRE MESH.
DUST SHIELDS HELPED KEEP THE ROVER MOVING.

The lunar rover
Apollo 15, *16*, and *17* each took a lunar rover to the Moon. It was folded up and kept in the storage bay of the lunar lander. On the Moon, astronauts unfolded it and drove it. They used it to travel much farther than they could possibly walk.

Space Shuttle

In the early days of space exploration, spacecraft were only able to fly one mission. Most of the rockets' parts burned up in Earth's atmosphere or were left in space. The space shuttle is the first spacecraft that can be used more than once. The first space shuttle, called *Enterprise*, never flew in space, but it proved that a craft the size of the shuttle could glide successfully down to Earth.

The second space shuttle, *Columbia*, was launched in 1981. Space shuttles *Challenger*, *Discovery*, *Atlantis*, and *Endeavor* also launched in the 1980s. In 1986, the *Challenger* exploded after takeoff, killing all seven astronauts onboard.

Apart from the *Challenger* disaster, space shuttles have worked well. They are being used to construct the International Space Station (*see pages 28–29*).

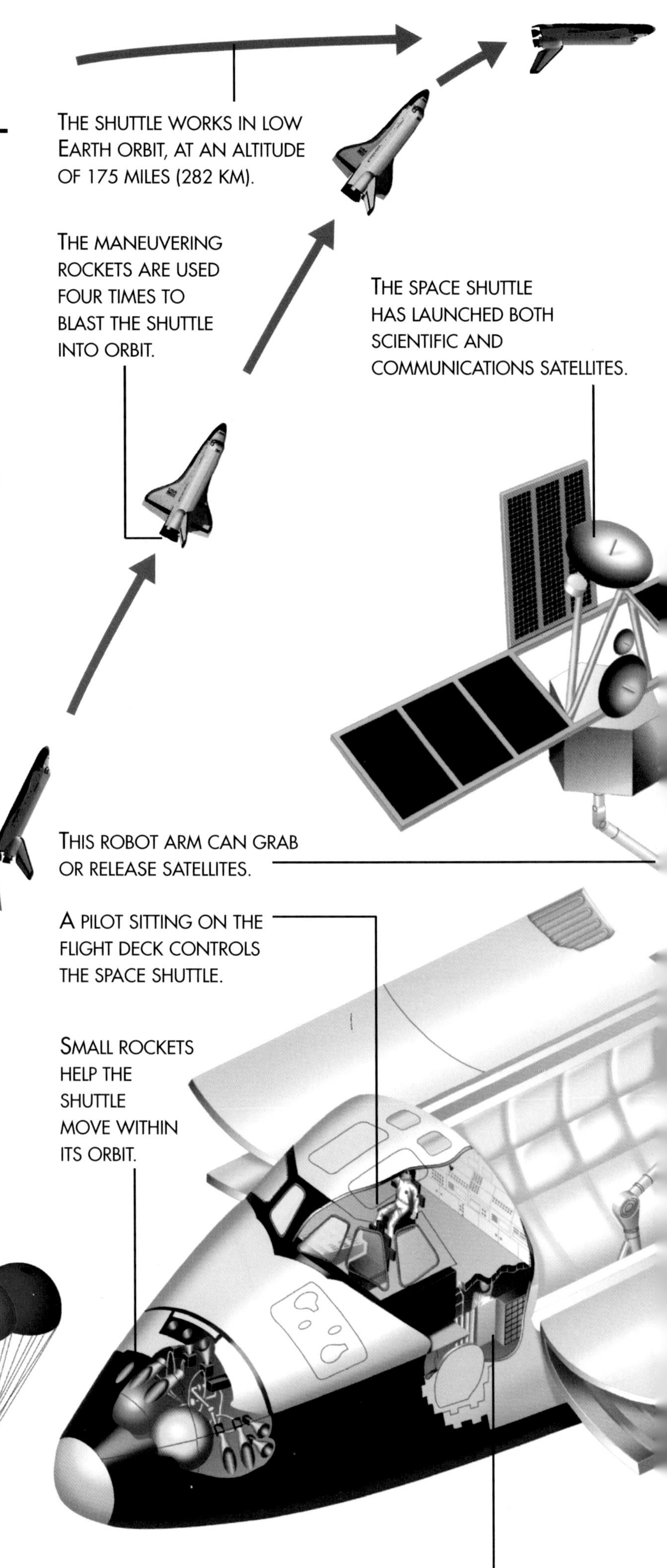

THE SHUTTLE WORKS IN LOW EARTH ORBIT, AT AN ALTITUDE OF 175 MILES (282 KM).
THE MANEUVERING ROCKETS ARE USED FOUR TIMES TO BLAST THE SHUTTLE INTO ORBIT.
THE SPACE SHUTTLE HAS LAUNCHED BOTH SCIENTIFIC AND COMMUNICATIONS SATELLITES.
THIS ROBOT ARM CAN GRAB OR RELEASE SATELLITES.
A PILOT SITTING ON THE FLIGHT DECK CONTROLS THE SPACE SHUTTLE.
SMALL ROCKETS HELP THE SHUTTLE MOVE WITHIN ITS ORBIT.
SOLID ROCKET BOOSTERS ARE REUSED FOR OTHER SHUTTLE FLIGHTS.
ASTRONAUTS LIVE AND WORK ON THE LOWER DECK DURING THE FLIGHT.

THE MAIN FUEL TANK IS NOT REUSED.

BOOSTER ROCKETS SEPARATE AND FALL INTO THE OCEAN.

Shuttle flight profile
The space shuttle blasts off from Cape Canaveral, Florida, in the United States. At a height, or altitude, of about 31 miles (50 km), the solid rocket boosters separate and fall back down to Earth. At an altitude of about 75 miles (120 km), the shuttle ejects the main fuel tank, which is now empty. The shuttle then moves fast enough to reach orbit.

AT LAUNCH, THE EXTERNAL FUEL TANK SUPPORTS THE BODY OF THE SHUTTLE.

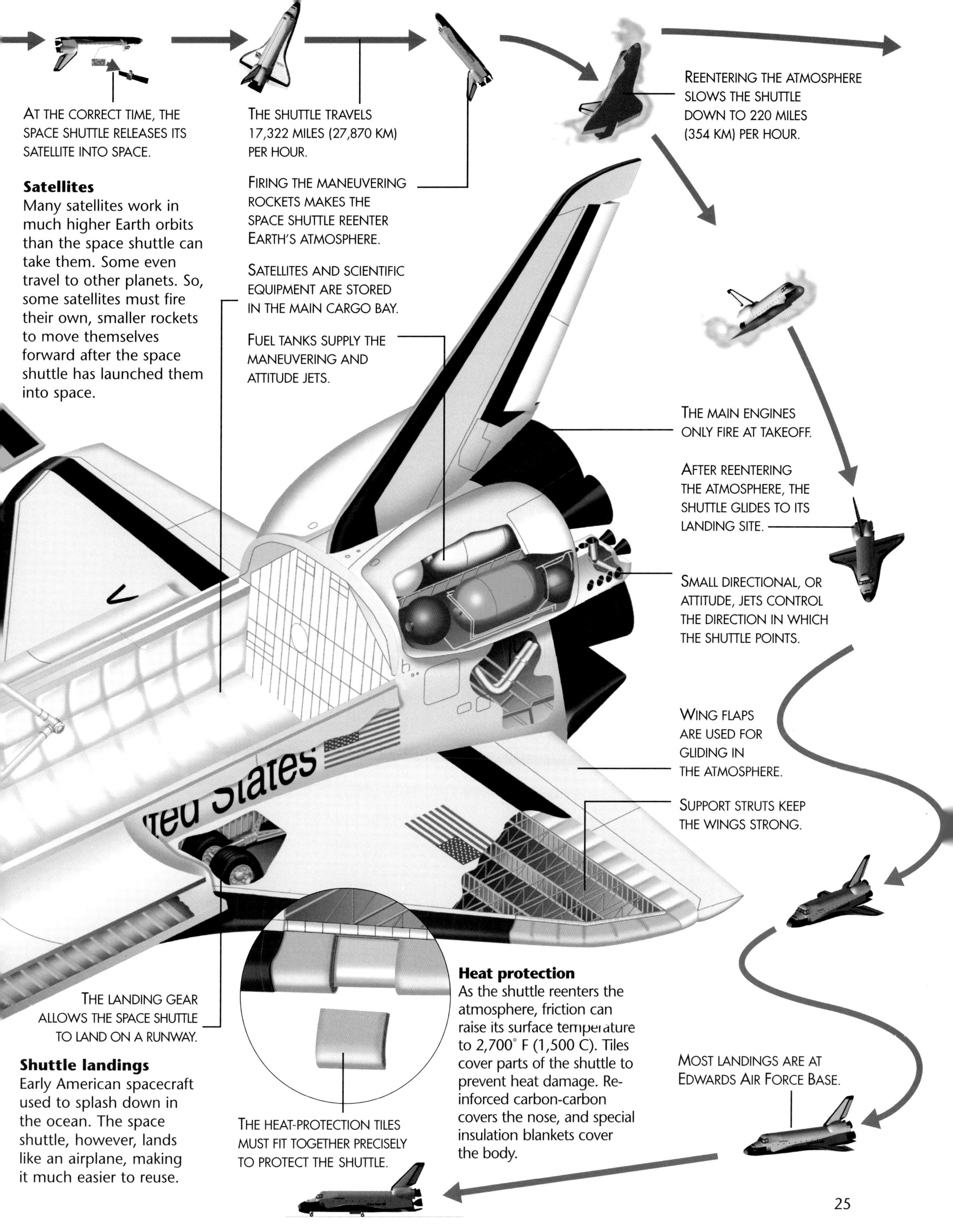
AT THE CORRECT TIME, THE SPACE SHUTTLE RELEASES ITS SATELLITE INTO SPACE.
THE SHUTTLE TRAVELS 17,322 MILES (27,870 KM) PER HOUR.
FIRING THE MANEUVERING ROCKETS MAKES THE SPACE SHUTTLE REENTER EARTH'S ATMOSPHERE.
REENTERING THE ATMOSPHERE SLOWS THE SHUTTLE DOWN TO 220 MILES (354 KM) PER HOUR.
SATELLITES AND SCIENTIFIC EQUIPMENT ARE STORED IN THE MAIN CARGO BAY.
FUEL TANKS SUPPLY THE MANEUVERING AND ATTITUDE JETS.
THE MAIN ENGINES ONLY FIRE AT TAKEOFF.
AFTER REENTERING THE ATMOSPHERE, THE SHUTTLE GLIDES TO ITS LANDING SITE.
SMALL DIRECTIONAL, OR ATTITUDE, JETS CONTROL THE DIRECTION IN WHICH THE SHUTTLE POINTS.
WING FLAPS ARE USED FOR GLIDING IN THE ATMOSPHERE.
SUPPORT STRUTS KEEP THE WINGS STRONG.
THE LANDING GEAR ALLOWS THE SPACE SHUTTLE TO LAND ON A RUNWAY.
THE HEAT-PROTECTION TILES MUST FIT TOGETHER PRECISELY TO PROTECT THE SHUTTLE.
MOST LANDINGS ARE AT EDWARDS AIR FORCE BASE.

Satellites
Many satellites work in much higher Earth orbits than the space shuttle can take them. Some even travel to other planets. So, some satellites must fire their own, smaller rockets to move themselves forward after the space shuttle has launched them into space.

Shuttle landings
Early American spacecraft used to splash down in the ocean. The space shuttle, however, lands like an airplane, making it much easier to reuse.

Heat protection
As the shuttle reenters the atmosphere, friction can raise its surface temperature to 2,700° F (1,500 C). Tiles cover parts of the shuttle to prevent heat damage. Reinforced carbon-carbon covers the nose, and special insulation blankets cover the body.

Space Walking

Space has no gases for humans to breathe. Whenever work needs to be done outside a spacecraft, an astronaut must wear a space suit. A space suit supplies an astronaut with oxygen to breathe and water to drink.

On Earth, the weight of atmospheric pressure prevents fluids in our bodies from turning into gases. In space, there is no atmosphere, so a space suit must squeeze an astronaut's body to keep body fluids liquid.

On Earth and in spacecraft, only one-fifth of the gas breathed is oxygen. In a space suit, all of the gas is oxygen. Before astronauts can go outside in space suits, they must spend several hours breathing pure oxygen through masks, so their bodies can adapt.

Attitude jets
The attitude jets push the space walker along by blowing out nitrogen gas. The astronaut decides which way to move and pulls or twists the control levers. A computer calculates which jet should be fired. To make an astronaut turn, for example, more than one jet is fired at the same time.

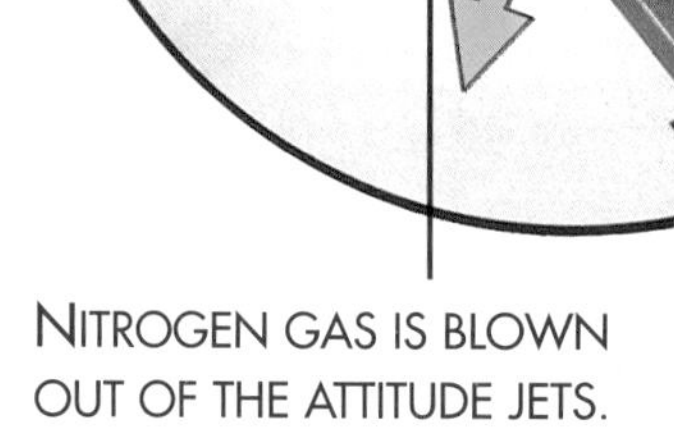

Moving in space
An astronaut has to wear a backpack to move around in space. This backpack has small jets, which are controlled by hand-held levers. The left-hand lever is used to make the space walker go forward or backward. The right-hand lever controls the direction the astronaut faces. It can make the space walker turn left or right, move up or down, or twist around clockwise or counterclockwise.

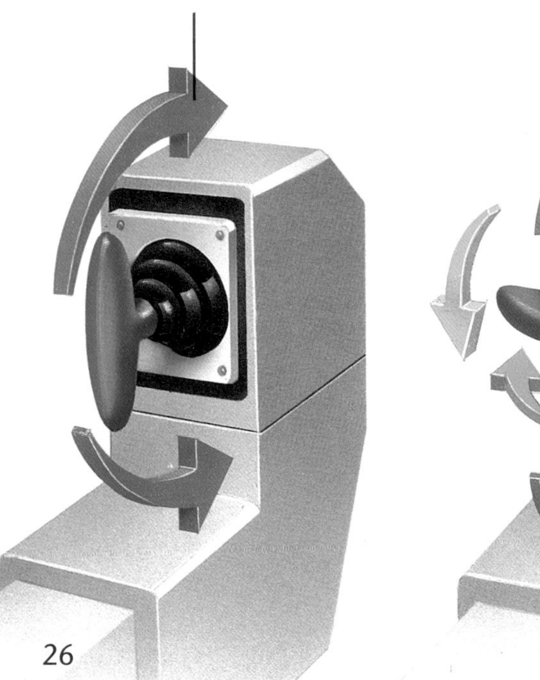
PUSH UP TO MOVE FORWARD.

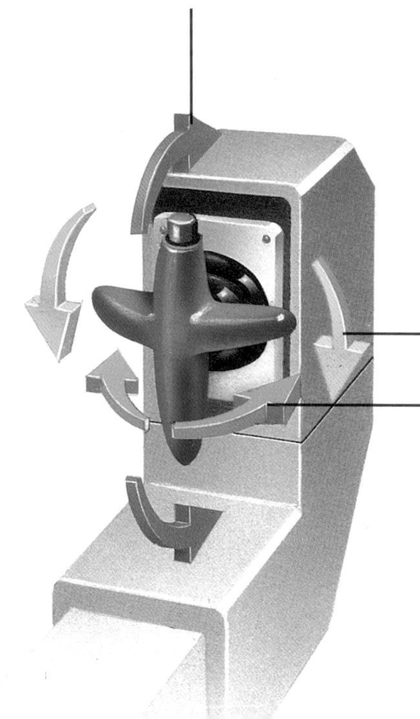
PUSH UP TO ROLL OVER.
TWISTING THE CONTROL LEVER MAKES THE ASTRONAUT ROLL TO THE LEFT OR RIGHT.
PULLING THE LEVER LEFT OR RIGHT MAKES THE ASTRONAUT TURN.

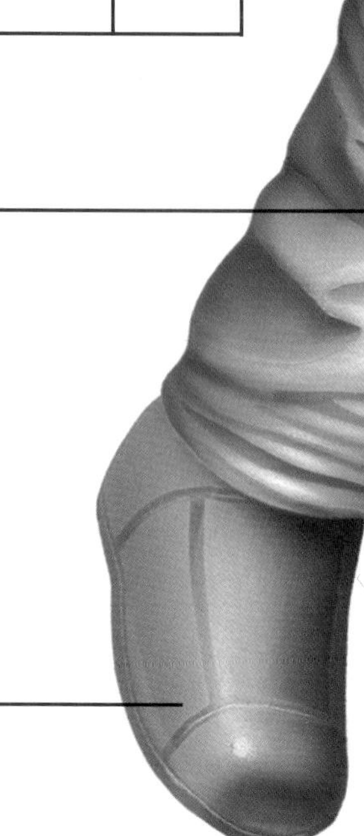
THIS VIDEO CAMERA ALLOWS THE ASTRONAUT TO RECORD IMAGES.
NITROGEN IS KEPT IN A TANK AND SUPPLIED TO THE ATTITUDE JETS WHEN NEEDED.
THE BACKPACK IS KNOWN AS THE MANNED MANEUVERING UNIT (MMU).
A BATTERY SUPPLIES ELECTRICAL POWER TO THE MMU.
NITROGEN GAS IS BLOWN OUT OF THE ATTITUDE JETS.
ATTITUDE JETS POINT OUT OF THE BACKPACK IN ALL DIRECTIONS.
THE SUN WILL HEAT THE SPACE SUIT TO MORE THAN 212° FAHRENHEIT (100° C).
BOOTS ARE ATTACHED TO THE SPACE SUIT TO PREVENT OXYGEN FROM ESCAPING.

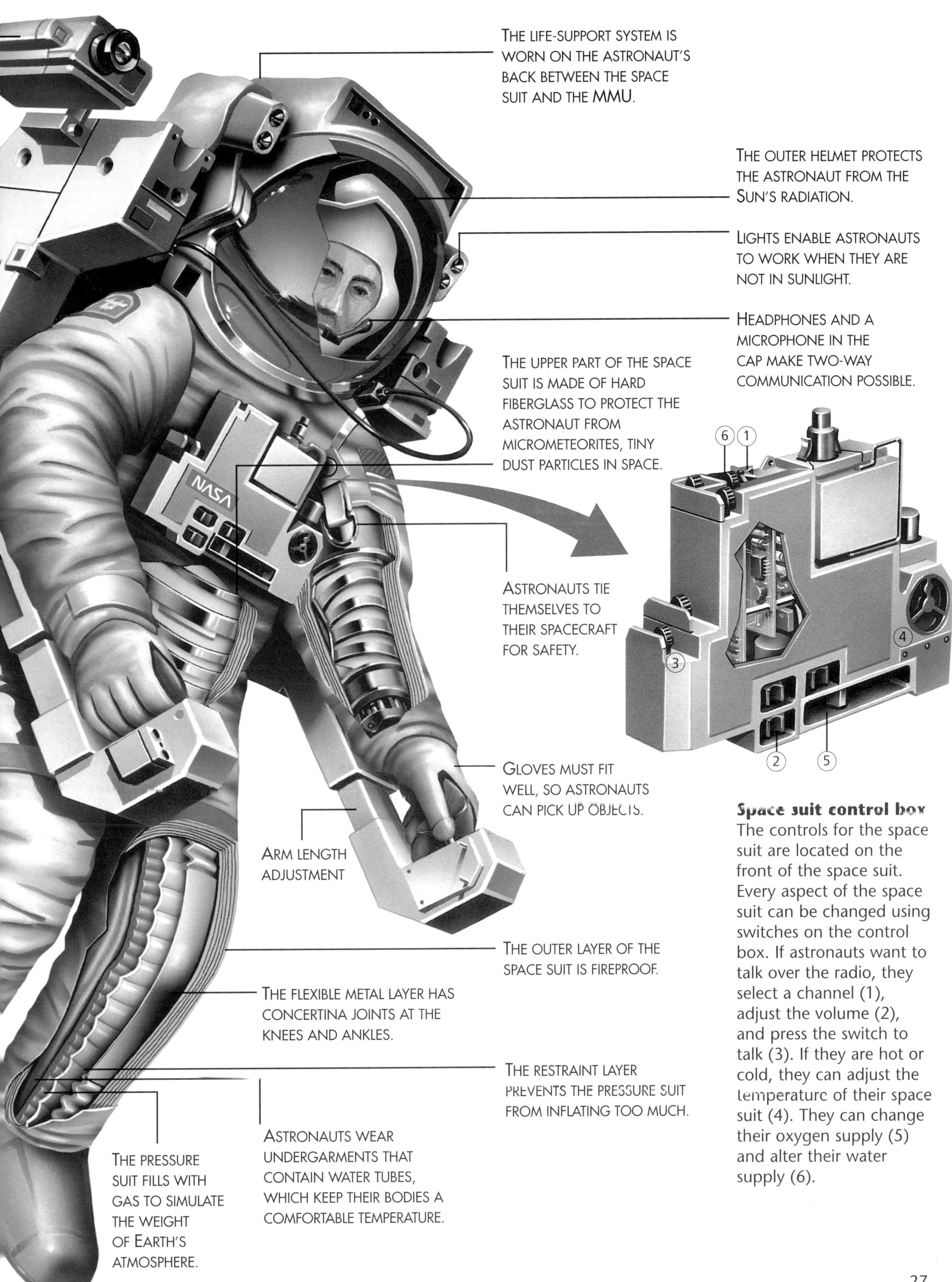
THE LIFE-SUPPORT SYSTEM IS WORN ON THE ASTRONAUT'S BACK BETWEEN THE SPACE SUIT AND THE MMU.
THE OUTER HELMET PROTECTS THE ASTRONAUT FROM THE SUN'S RADIATION.
LIGHTS ENABLE ASTRONAUTS TO WORK WHEN THEY ARE NOT IN SUNLIGHT.
HEADPHONES AND A MICROPHONE IN THE CAP MAKE TWO-WAY COMMUNICATION POSSIBLE.
THE UPPER PART OF THE SPACE SUIT IS MADE OF HARD FIBERGLASS TO PROTECT THE ASTRONAUT FROM MICROMETEORITES, TINY DUST PARTICLES IN SPACE.
NASA
6
1
ASTRONAUTS TIE THEMSELVES TO THEIR SPACECRAFT FOR SAFETY.
4
3
2
5
GLOVES MUST FIT WELL, SO ASTRONAUTS CAN PICK UP OBJECTS.
ARM LENGTH ADJUSTMENT
THE OUTER LAYER OF THE SPACE SUIT IS FIREPROOF.
THE FLEXIBLE METAL LAYER HAS CONCERTINA JOINTS AT THE KNEES AND ANKLES.
THE RESTRAINT LAYER PREVENTS THE PRESSURE SUIT FROM INFLATING TOO MUCH.
ASTRONAUTS WEAR UNDERGARMENTS THAT CONTAIN WATER TUBES, WHICH KEEP THEIR BODIES A COMFORTABLE TEMPERATURE.
THE PRESSURE SUIT FILLS WITH GAS TO SIMULATE THE WEIGHT OF EARTH'S ATMOSPHERE.

Space suit control box
The controls for the space suit are located on the front of the space suit. Every aspect of the space suit can be changed using switches on the control box. If astronauts want to talk over the radio, they select a channel (1), adjust the volume (2), and press the switch to talk (3). If they are hot or cold, they can adjust the temperature of their space suit (4). They can change their oxygen supply (5) and alter their water supply (6).

Space Stations

In 1971, the Soviet Union launched the first space station, *Salyut 1*. Early space stations only stayed in orbit for a few months before falling back to Earth. Gradually, better systems were developed. In 1986, the Soviet Union launched the space station *Mir*, which is still in orbit today.

The U.S. launched the *Skylab* space station in 1973. Astronauts worked there for 172 days before returning. *Skylab* burned up in Earth's atmosphere before it was revisited.

The International Space Station (*right*) is the largest space station ever planned. A group of sixteen countries, including the United States, Russia, Canada, and Japan, is working to complete the more than thirty flights needed to take all the pieces into space. Nearly forty other missions will be needed to put the space station together.

Space station construction
The International Space Station will take several years to construct. The first stage will be the delivery of the central connecting structure (1). During the second stage, three astronauts will board the station (2). In the third stage, many of the scientific modules will be added (3). When the station is completed, in 2004, it will always be staffed by astronauts (4) from around the world.

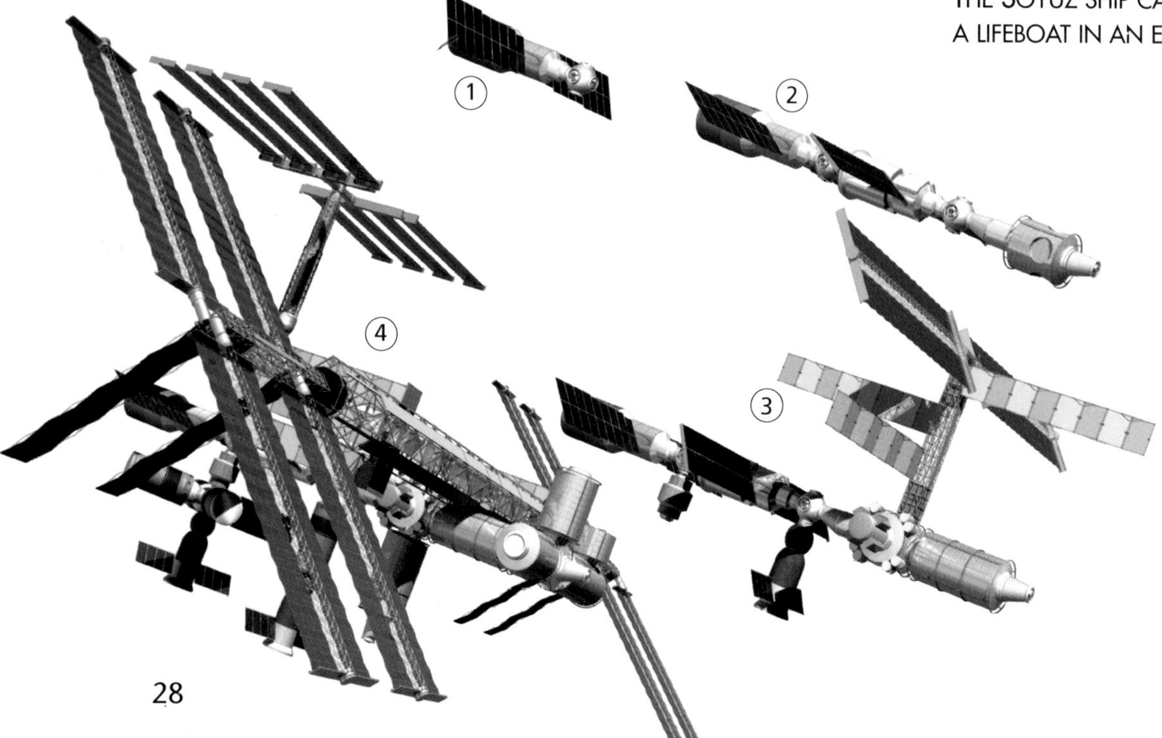
1
2
3
4

STRONG LATTICE FRAMEWORKS WILL SUPPORT THE SPACE STATION.

SOLAR PANELS WILL GENERATE POWER.

NEW MODULES CAN BE FITTED TO THE STATION AT SPECIAL CONNECTION POINTS.

RUSSIA'S SOYUZ SPACECRAFT WILL REMAIN PERMANENTLY ATTACHED TO THE SPACE STATION.

THE SOYUZ SHIP CAN ACT AS A LIFEBOAT IN AN EMERGENCY.

AN ADDITIONAL SOYUZ SPACECRAFT WILL BE USED FOR TRANSFERRING CREW.

Habitation module
The habitation module is one of the most important parts of the space station because it is where the astronauts will live and sleep. Astronauts have to sleep with fans close to their faces. The fans blow away the carbon dioxide they breathe out.

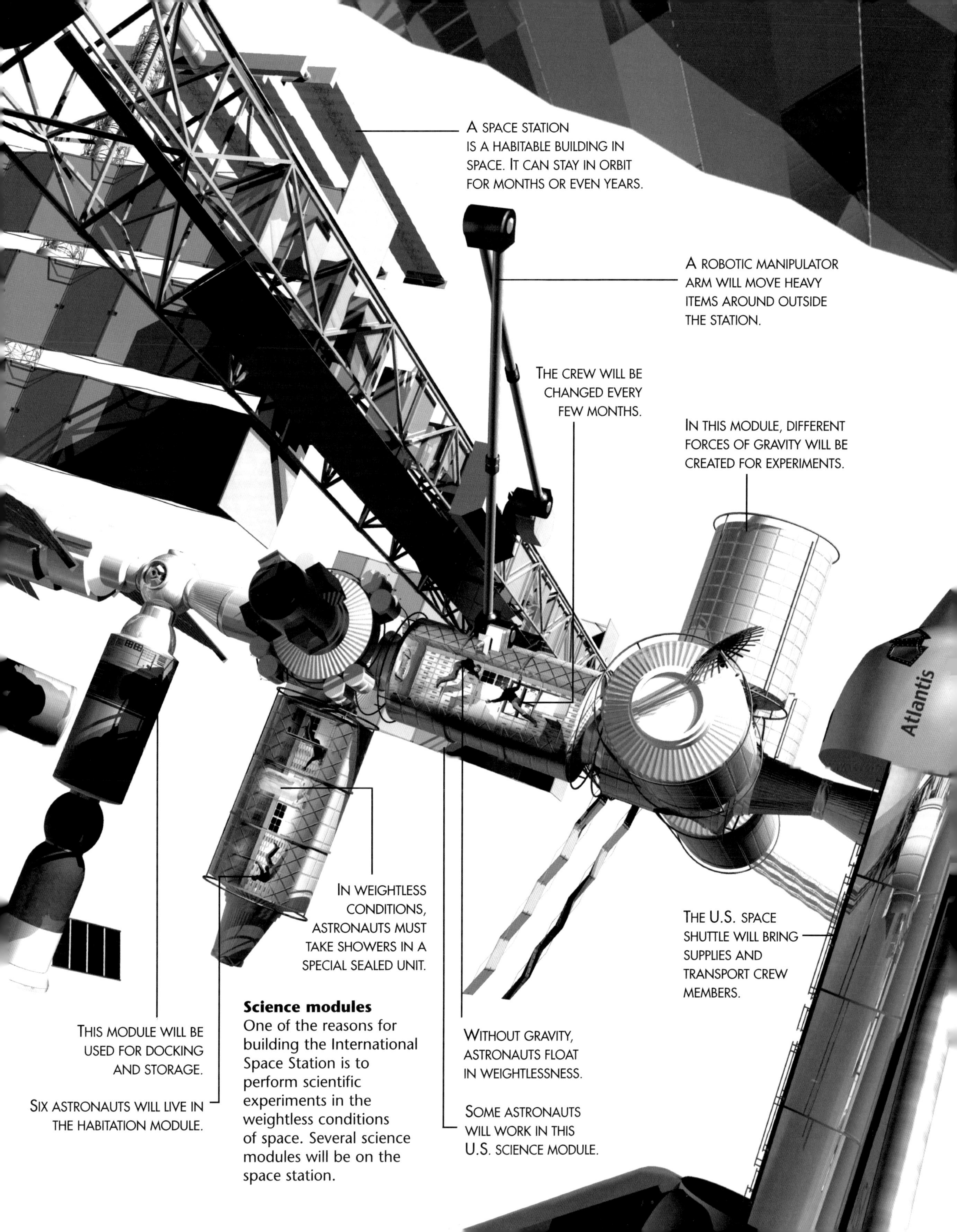
A SPACE STATION IS A HABITABLE BUILDING IN SPACE. IT CAN STAY IN ORBIT FOR MONTHS OR EVEN YEARS.
A ROBOTIC MANIPULATOR ARM WILL MOVE HEAVY ITEMS AROUND OUTSIDE THE STATION.
THE CREW WILL BE CHANGED EVERY FEW MONTHS.
IN THIS MODULE, DIFFERENT FORCES OF GRAVITY WILL BE CREATED FOR EXPERIMENTS.
Atlantis
IN WEIGHTLESS CONDITIONS, ASTRONAUTS MUST TAKE SHOWERS IN A SPECIAL SEALED UNIT.
THE U.S. SPACE SHUTTLE WILL BRING SUPPLIES AND TRANSPORT CREW MEMBERS.
THIS MODULE WILL BE USED FOR DOCKING AND STORAGE.
SIX ASTRONAUTS WILL LIVE IN THE HABITATION MODULE.
WITHOUT GRAVITY, ASTRONAUTS FLOAT IN WEIGHTLESSNESS.
SOME ASTRONAUTS WILL WORK IN THIS U.S. SCIENCE MODULE.

Science modules
One of the reasons for building the International Space Station is to perform scientific experiments in the weightless conditions of space. Several science modules will be on the space station.

Viking Landers

The *Viking* missions were the first to place working landers on the surface of Mars. Scientists began the project in 1968. They built two *Viking* spacecraft because they hoped one would continue working even if the other broke down. In fact, both space probes worked very well. The landers touched down on Mars in the summer of 1976 and worked for many years.

Along with the landers, each of the *Viking* space probes had a section that stayed in orbit. The two *Viking* orbiters surveyed and mapped nearly all of the surface of Mars.

Many scientists think that Mars was once much like Earth. There is evidence that parts of its surface were once covered with water. Some scientists also believe that there may have been life-forms on Mars. The *Viking* landers had special equipment onboard to test the soil for tiny living organisms called microbes. However, none were found.

The *Viking* landings
The site of *Viking I's* touchdown was Chryse Planitia, close to the Martian equator. *Viking II* landed farther north in Utopia Planitia. Both places are flat and desertlike. Heavily cratered areas were avoided since they might have damaged the landing spacecraft.

THIS ANTENNA SENT INFORMATION TO EARTH AND RECEIVED COMMANDS.

CAMERAS OBSERVED A TEST PATTERN TO GET THEIR COLOR BALANCE CORRECT.

MAGNETS ON THE TEST CHART WERE SENSITIVE TO MAGNETIC ACTIVITY ON MARS.

THIS MACHINE TESTED FOR EARTHQUAKES ON MARS AND FOUND NONE.
VIKING USED TWO CAMERAS TO PHOTOGRAPH THE MARTIAN SURFACE.
POWER GENERATORS
FUEL TANK UNDER WINDSHIELD

MARTIAN SOIL IS RED BECAUSE IT IS COVERED WITH RUST.

Flight path
It took the *Viking* space probes almost one year to reach Mars. Once in the planet's orbit, the probes photographed the surface and sent the images to Earth. Scientists used these images to pick landing sites. When they chose a site, the lander was launched from the orbiter. A parachute attached to the lander slowed it down and helped it land gently.

THE BIOLOGY PROCESSOR CHECKED THE RESULTS OF EXPERIMENTS LOOKING FOR MARTIAN LIFE.
THIS DEVICE HEATED MARTIAN SOIL TO TEST FOR MICROBES.
THE METEOROLOGY ARM CHECKED THE WEATHER IN THE AREA WHERE VIKING LANDED.
SHOCK ABSORBERS ON THE LANDING LEGS PREVENTED VIKING FROM BEING DAMAGED DURING TOUCHDOWN.
THE ROBOTIC ARM ON VIKING I WOULD NOT WORK PROPERLY IN THE COLD, MARTIAN MORNING.
MARS IS COVERED WITH SOIL AND ROCKS.
THE DESCENT ENGINE HELPED THE PROBE LAND.

Pictures of Mars
Viking I began transmitting photographs to Earth just twenty-five seconds after it touched down on Mars. Because of the long distance from Mars to Earth, the images took nineteen minutes to arrive on radio waves. One picture (*above*) showed a large rock. Scientists named it Big Joe. Fewer big rocks existed in the region where *Viking II* landed.

The robotic arm
Viking was designed to test the surface of Mars for signs of life. It did this by scooping soil from the surface and dropping it into a biology processor and heating device. Commands were sent from Earth to the robotic arm, which performed them automatically.

Mars *Pathfinder*

The spacecraft *Pathfinder* landed on Mars on July 4, 1997. It was the first spacecraft to land on the planet since *Vikings 1* and *2*. *Pathfinder* searched for clues to learn about water and rocks on Mars. Mars has features on its surface that only flowing water could make. Yet, no liquid water now exists on the planet. Scientists want to learn what happened to the water.

Weather conditions were harsh when *Pathfinder* landed on Mars, and scientists worried that its instruments might fail. *Pathfinder*, however, worked well and sent data back to Earth for weeks. It finally stopped working on September 27, 1997.

Pathfinder was the first in a series of space probes scientists plan to send to Mars. After *Pathfinder*, scientists sent the probe *Mars Global Surveyor* to orbit and map the planet. Additional probes will have tiny rovers that will drive over the surface of Mars to look for signs of water.

Mars landscape
Scientists chose an ancient floodplain as the landing site for *Pathfinder*. Scientists estimate that this flood took place several billion years ago. The surface of Mars has remained almost unchanged since then. *Pathfinder* came to rest in a small dip.

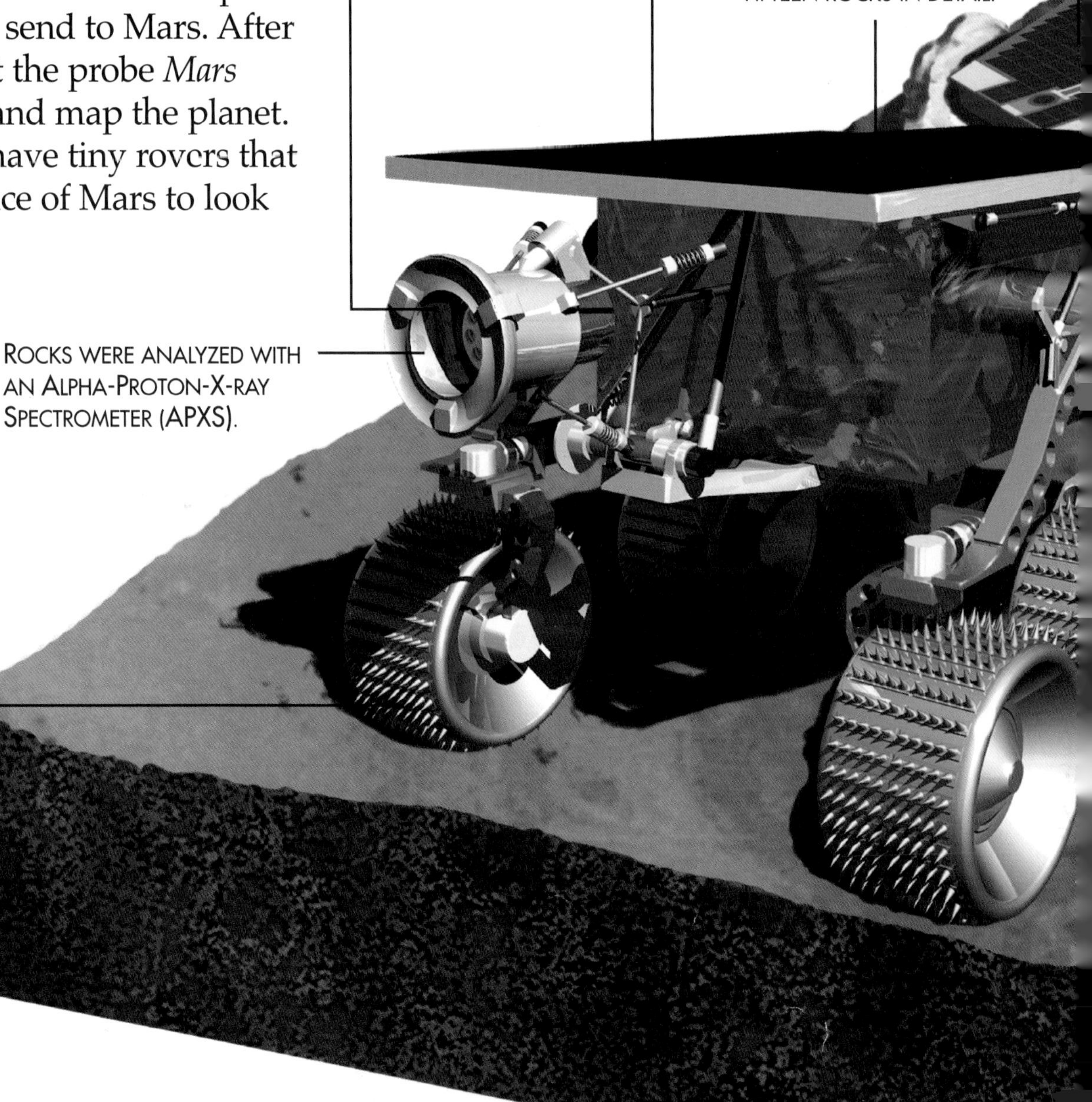

A LARGE ANTENNA WAS USED TO COMMUNICATE WITH EARTH.

THE APXS INSTRUMENT FIRED SUB-ATOMIC PARTICLES AT THE ROCKS TO IDENTIFY THEIR CHEMICAL COMPOSITION.

SOLAR PANELS COLLECTED ENERGY.

THIS SOLAR PANEL ON TOP OF SOJOURNER COLLECTED ENERGY.

THE ROVER ANALYZED FIFTEEN ROCKS IN DETAIL.

ROCKS WERE ANALYZED WITH AN ALPHA-PROTON-X-RAY SPECTROMETER (APXS).

THE WHEELS ON SOJOURNER WERE SPIKED SO THEY COULD GRIP THE SOIL.

Mars rover
The rover on the Mars *Pathfinder* was called *Sojourner*. It drove across the surface of the planet and went up close to many of the rocks near *Pathfinder*. Once close to a rock, it performed tests to analyze the rocks' mineral content.

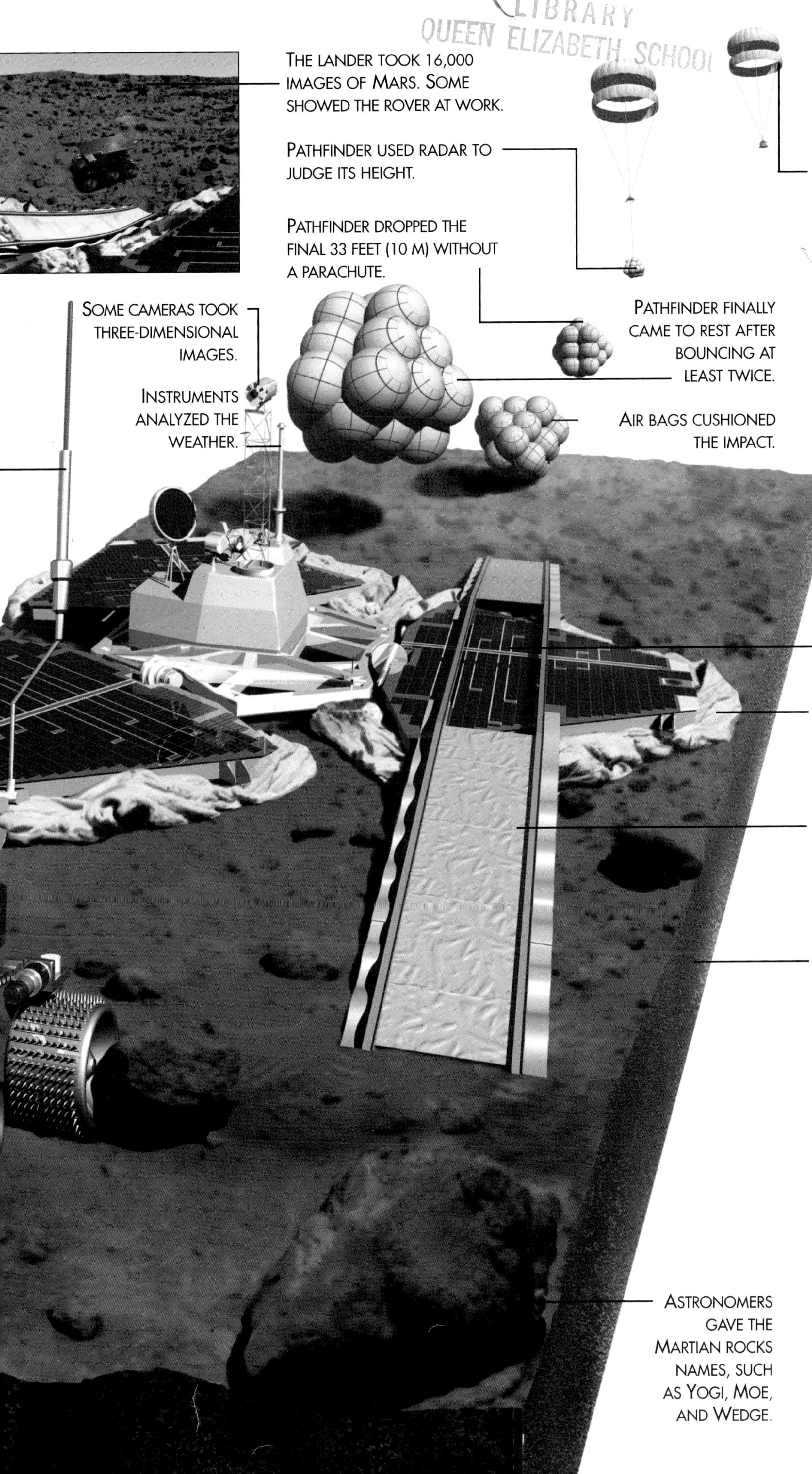
THE LANDER TOOK 16,000 IMAGES OF MARS. SOME SHOWED THE ROVER AT WORK.
PATHFINDER USED RADAR TO JUDGE ITS HEIGHT.
PATHFINDER DROPPED THE FINAL 33 FEET (10 M) WITHOUT A PARACHUTE.
PATHFINDER ENTERED THE ATMOSPHERE 78 MILES (125 KM) ABOVE MARS.
ITS PARACHUTE OPENED 6 MILES (10 KM) ABOVE THE SURFACE.
SOME CAMERAS TOOK THREE-DIMENSIONAL IMAGES.
INSTRUMENTS ANALYZED THE WEATHER.
PATHFINDER FINALLY CAME TO REST AFTER BOUNCING AT LEAST TWICE.
AIR BAGS CUSHIONED THE IMPACT.
SOJOURNER WAS STORED ON ONE OF THREE "PETALS." AFTER LANDING, THE PETALS OPENED.
AFTER PATHFINDER CAME TO REST, ITS AIR BAGS DEFLATED.
A RAMP UNROLLED FROM A PETAL, AND SOJOURNER DROVE ONTO THE MARTIAN SURFACE.
WATER MAY BE FOUND IN A PERMANENT FROST LAYER UNDER THE SURFACE OF MARS.
ASTRONOMERS GAVE THE MARTIAN ROCKS NAMES, SUCH AS YOGI, MOE, AND WEDGE.

Landing on Mars
Pathfinder was designed well. Instead of carrying bulky rockets to slow it down for landing, it was fitted with large air bags. These inflated as *Pathfinder* neared the surface, and they stopped any damage from occurring when the spacecraft struck the surface of Mars.

The Mars lander
Pathfinder was split into two spacecraft, the lander and the rover. The lander took readings of Martian weather conditions, such as temperatures and wind speeds. It stayed in touch with Earth and relayed messages to the rover. Shortly after it landed, *Pathfinder* was renamed the *Sagan Memorial Station* in honor of the famous U.S. scientist Carl Sagan, who had just died.

Voyager

Space probes *Voyagers 1* and *2* flew two of the most successful space missions in history. *Voyager 2* was launched on August 20, 1977, and *Voyager 1* was launched on September 5, 1977. *Voyager 1* followed a faster route to Jupiter and passed *Voyager 2* along the way.

In March 1979, *Voyager 1* flew past Jupiter and sent scientific information and images of the planet to Earth. It continued its mission, flying by Saturn and its largest moon, Titan, in November 1980.

Voyager 2 reached Jupiter in July 1979 and Saturn in 1981. Unlike *Voyager 1*, it did not fly close to Titan. Instead, it went on to Uranus. In January 1986, it transmitted pictures of this pale-blue planet to Earth. It then began its journey to Neptune. In August 1989, *Voyager 2* reached Neptune, the final planet on its mission. Only the distant planet Pluto remains to be visited by a space probe.

The grand tour
The *Voyager* space probes followed gravity-assisted orbits. These orbits were first planned in 1965 by Gary Flandro, who realized that the four gas giants line up only once every 175 years. This meant that a space probe could visit each giant planet by using the gravity of one to throw it onward to the next. He calculated that the planets would next line up in the 1980s, and so the idea for the *Voyager* missions was born.

VOYAGER'S GOLDEN DISK PLAYS EARTH SOUNDS, SUCH AS THE WORD "HELLO" SPOKEN IN FIFTY-FIVE LANGUAGES.

THE MAGNETOMETER BOOM CONTAINS DETECTORS THAT MEASURE MAGNETIC FIELDS IN SPACE.

The *Voyagers* today
On February 17, 1998, *Voyager 1* became the most distant object made by humans when it cruised out of the Solar System. *Voyager 1* is traveling above the Solar System. Eventually, *Voyager 2* will travel below the Solar System.

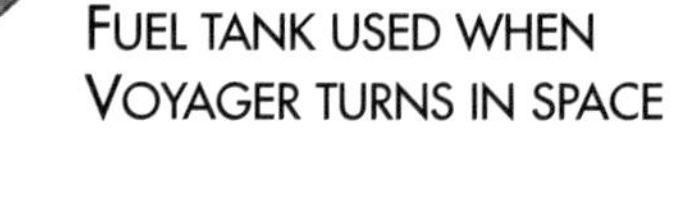

FUEL TANK USED WHEN VOYAGER TURNS IN SPACE

THE SUN IS TOO FAINT TO BE AN ENERGY SOURCE IN THE OUTER SOLAR SYSTEM, SO THREE RADIOISOTOPE THERMOELECTRIC GENERATORS PROVIDE VOYAGER WITH ENERGY.

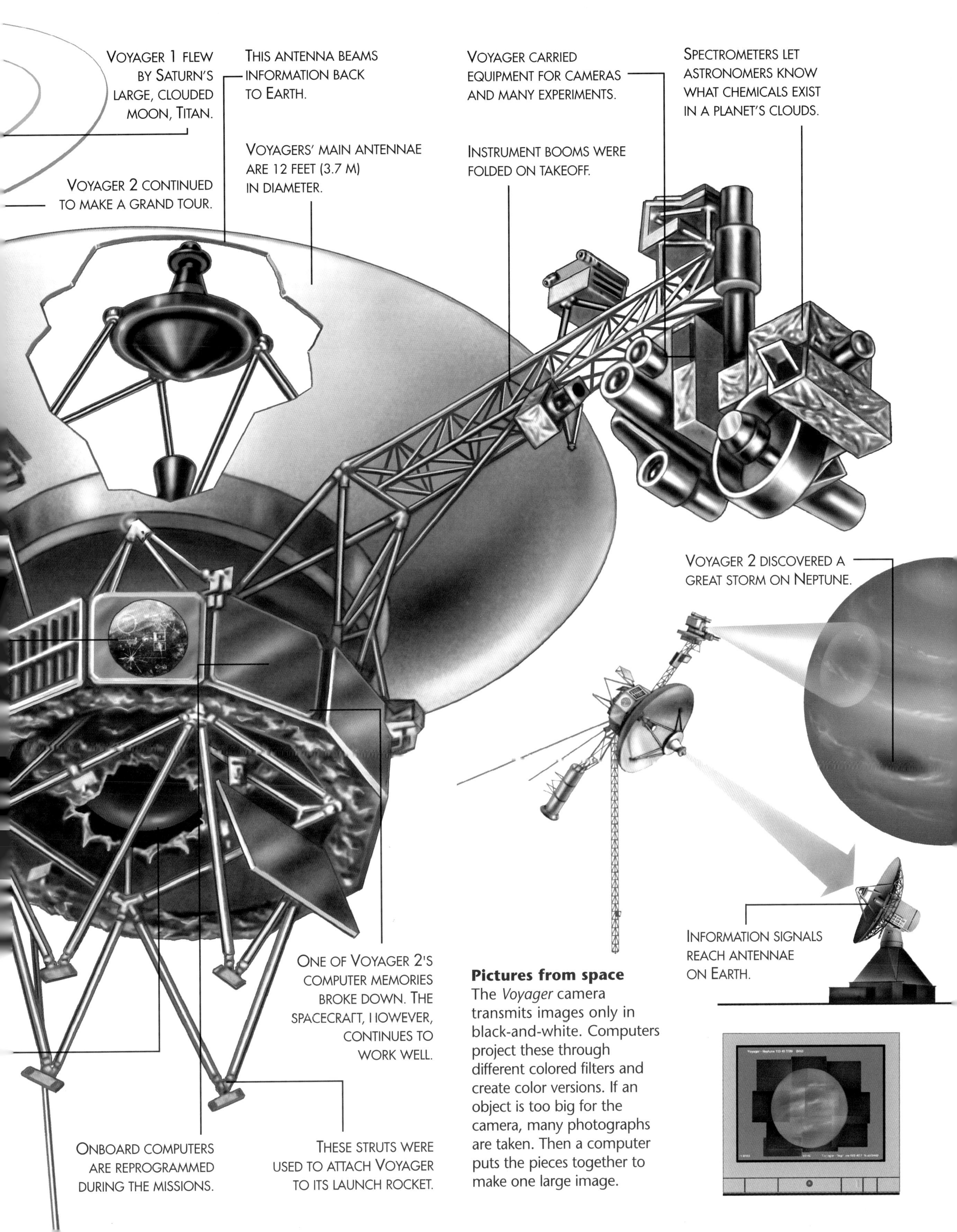
Voyager 1 flew by Saturn's large, clouded moon, Titan.
This antenna beams information back to Earth.
Voyager carried equipment for cameras and many experiments.
Spectrometers let astronomers know what chemicals exist in a planet's clouds.
Voyagers' main antennae are 12 feet (3.7 m) in diameter.
Instrument booms were folded on takeoff.
Voyager 2 continued to make a grand tour.
Voyager 2 discovered a great storm on Neptune.
Information signals reach antennae on Earth.
One of Voyager 2's computer memories broke down. The spacecraft, however, continues to work well.
Onboard computers are reprogrammed during the missions.
These struts were used to attach Voyager to its launch rocket.

Pictures from space

The *Voyager* camera transmits images only in black-and-white. Computers project these through different colored filters and create color versions. If an object is too big for the camera, many photographs are taken. Then a computer puts the pieces together to make one large image.

Studying Stars

Astronomers use very large telescopes in observatories (*right*) to study the stars. These telescopes make faraway objects appear larger and closer by using two mirrors to focus light. Like any piece of machinery, a telescope sometimes needs to be repaired, so everything necessary to keep it working is kept in an observatory.

A telescope operator sits in a control room and uses computer-controlled motors to move the telescope. Another astronomer collects data. Astronomers record their observations with detectors that consist of a variety of cameras and instruments. The detectors are attached to the telescope's base inside a metal cage. Data is fed into the control room, and then images appear on a computer screen. The images are stored on computer disks. Scientists from around the world carefully study the images and information on the computer disks.

Taking images
Electronic cameras take photographs of the night sky and display them on computers. Each picture is made up of tiny dots known as picture elements, or pixels. The image below shows solar systems forming in a cloud of gas known as the Orion nebula. They appear so small on the image that when the picture is enlarged, the square pixels can be seen.

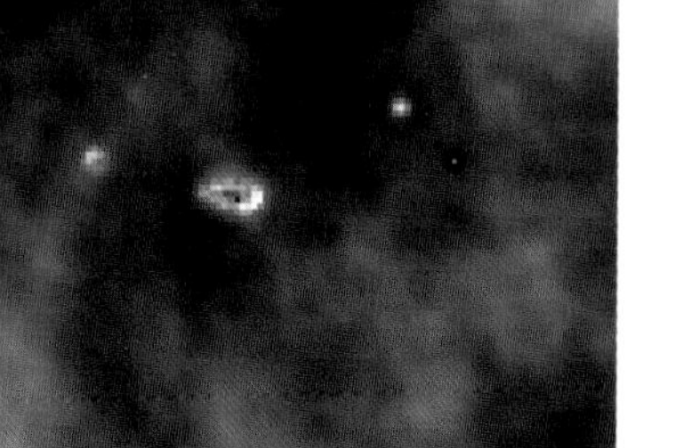

This opening allows the telescope to see out.

The front end of this reflecting telescope contains the secondary mirror.

Rotating the dome exposes a different part of the sky to the telescope.

A metal cage houses detectors attached to the telescope.

The telescope operator in the control room moves the telescope. Astronomers also sit here.

The ventilation system keeps the temperature inside the observatory at a comfortable level.

Instruments are built and repaired in the electronics laboratory.

Visitors ride an elevator to reach the telescope.

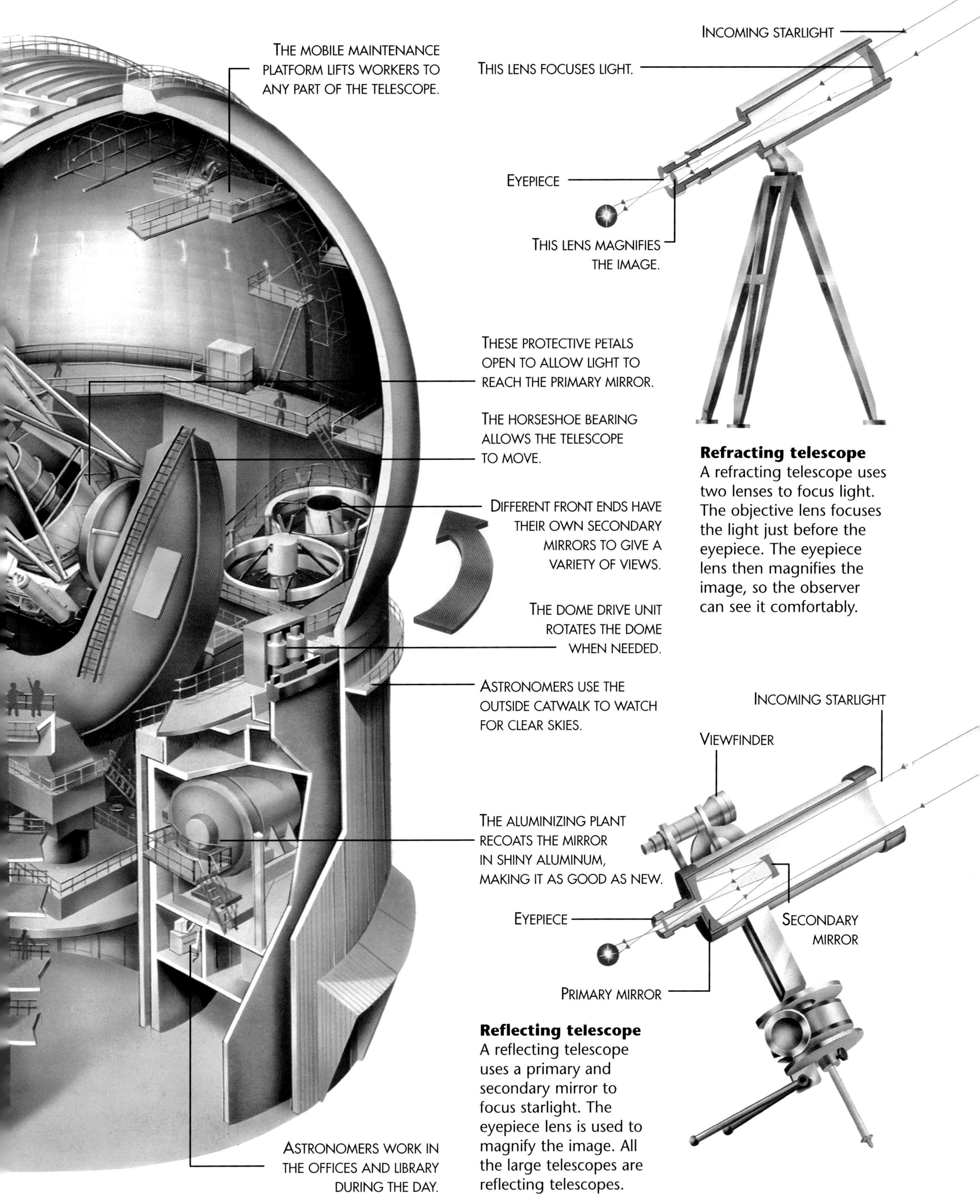
THE MOBILE MAINTENANCE PLATFORM LIFTS WORKERS TO ANY PART OF THE TELESCOPE.
THESE PROTECTIVE PETALS OPEN TO ALLOW LIGHT TO REACH THE PRIMARY MIRROR.
THE HORSESHOE BEARING ALLOWS THE TELESCOPE TO MOVE.
DIFFERENT FRONT ENDS HAVE THEIR OWN SECONDARY MIRRORS TO GIVE A VARIETY OF VIEWS.
THE DOME DRIVE UNIT ROTATES THE DOME WHEN NEEDED.
ASTRONOMERS USE THE OUTSIDE CATWALK TO WATCH FOR CLEAR SKIES.
THE ALUMINIZING PLANT RECOATS THE MIRROR IN SHINY ALUMINUM, MAKING IT AS GOOD AS NEW.
ASTRONOMERS WORK IN THE OFFICES AND LIBRARY DURING THE DAY.
INCOMING STARLIGHT
THIS LENS FOCUSES LIGHT.
EYEPIECE
THIS LENS MAGNIFIES THE IMAGE.
INCOMING STARLIGHT
VIEWFINDER
EYEPIECE
SECONDARY MIRROR
PRIMARY MIRROR

Refracting telescope
A refracting telescope uses two lenses to focus light. The objective lens focuses the light just before the eyepiece. The eyepiece lens then magnifies the image, so the observer can see it comfortably.

Reflecting telescope
A reflecting telescope uses a primary and secondary mirror to focus starlight. The eyepiece lens is used to magnify the image. All the large telescopes are reflecting telescopes.

Stars

Stars are enormous balls of hot gas that give off heat and light. Stars form inside collapsing clouds of hydrogen gas. The gas is squeezed so tightly that its temperature rises. When the temperature in the center of a star reaches 18 million° Fahrenheit (10 million° C), the hydrogen atoms' nuclei collide with such force that they stick together and form new helium nuclei. This nuclear fusion releases energy, which makes the star shine.

Stars can be many sizes. The Sun is a medium-sized star. Some stars have one hundred times the mass of the Sun. Stars that contain the most mass do not live as long as stars with little mass. A star with a very high mass will only live a few million years. A low-mass star, such as the Sun, could last 9 billion years or more.

The life of a star
Stars begin as protostars in the center of collapsing gas clouds (1). When the core of a protostar is hot enough, hydrogen begins to fuse into helium, and the protostar becomes a star (2). For most of its life, a star will generate

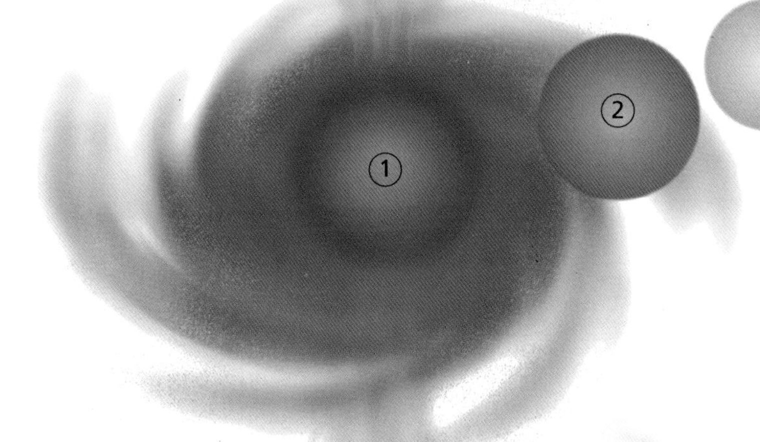

1
2

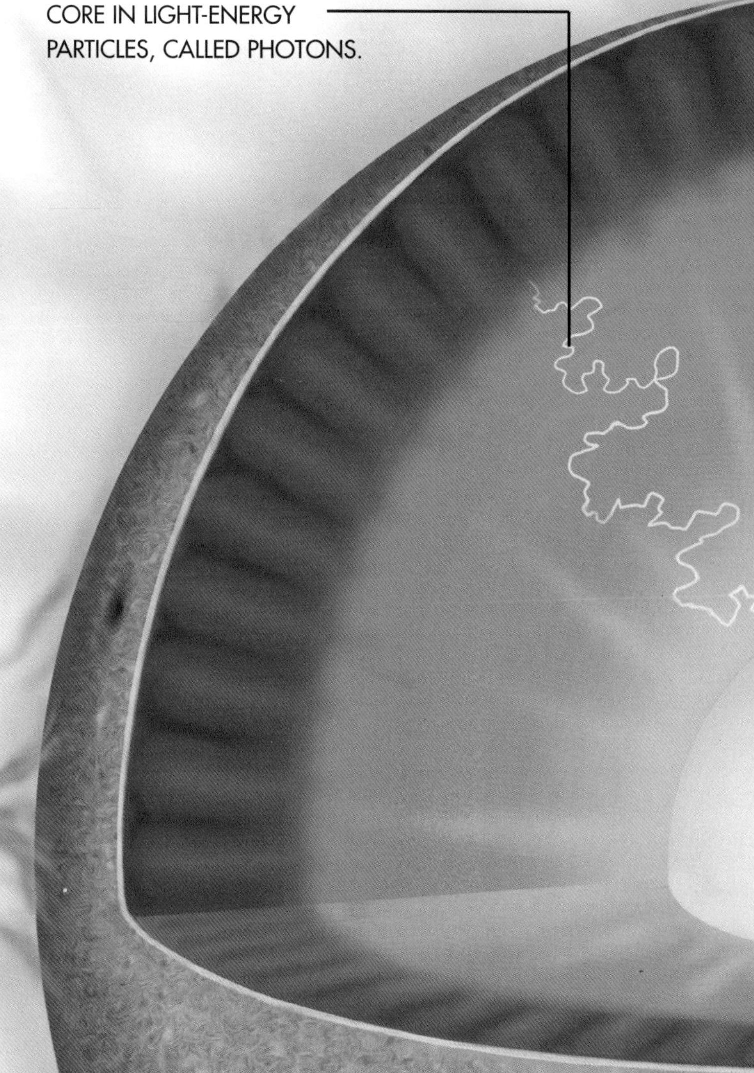

THE CORE OF THE SUN CONTAINS ABOUT ONE-TENTH OF THE SUN'S MASS.
ENERGY LEAVES THE SUN'S CORE IN LIGHT-ENERGY PARTICLES, CALLED PHOTONS.
ERUPTIVE PROMINENCES THROW SUPERHEATED GAS INTO SPACE.
THE SUN EMITS TINY PARTICLES THAT STREAM THROUGH SPACE. THESE PARTICLES MAKE UP SOLAR WIND.

The corona
The corona is the outermost gas around the Sun. It spreads very thinly throughout space and usually cannot be seen because the photosphere of the Sun is so bright. During a solar eclipse, the Moon blocks light from the photosphere, and the dimmer corona can be seen.

energy and change very little (3). As a star ages, the energy it generates increases (4), and it grows to become either a red giant or an exploding supernova. Late in a star's life, it may start to expand and contract, or pulsate (5). Nuclear fusion takes place in bursts rather than continuously. This unstable nuclear fusion lifts the star's outer layers and throws them off into space. That outer portion is called a planetary nebula (6) because to 1700s astronomers, these objects looked like planets.

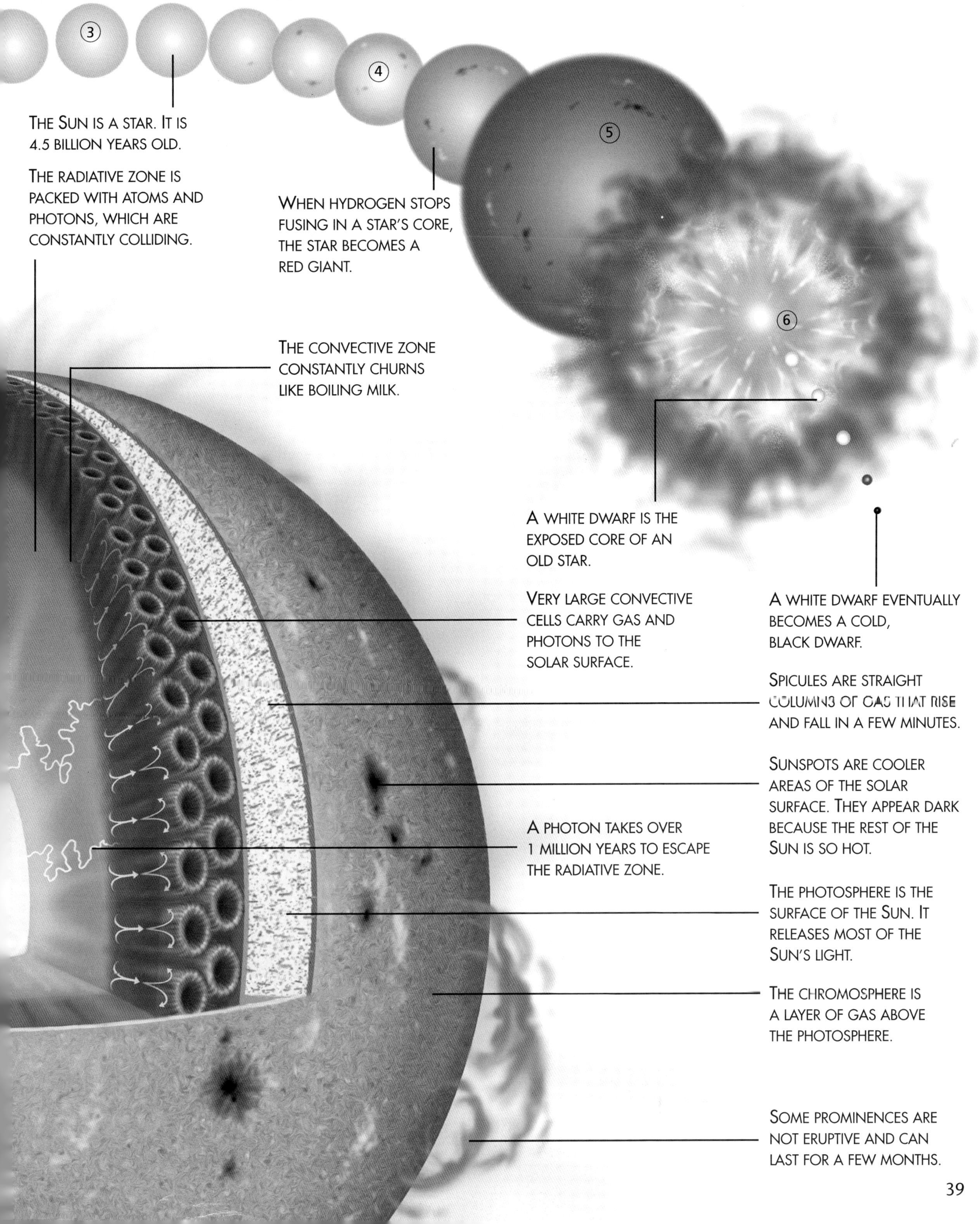

3

4

5

6

THE SUN IS A STAR. IT IS 4.5 BILLION YEARS OLD.

THE RADIATIVE ZONE IS PACKED WITH ATOMS AND PHOTONS, WHICH ARE CONSTANTLY COLLIDING.

WHEN HYDROGEN STOPS FUSING IN A STAR'S CORE, THE STAR BECOMES A RED GIANT.

THE CONVECTIVE ZONE CONSTANTLY CHURNS LIKE BOILING MILK.

A WHITE DWARF IS THE EXPOSED CORE OF AN OLD STAR.

VERY LARGE CONVECTIVE CELLS CARRY GAS AND PHOTONS TO THE SOLAR SURFACE.

A WHITE DWARF EVENTUALLY BECOMES A COLD, BLACK DWARF.

SPICULES ARE STRAIGHT COLUMNS OF GAS THAT RISE AND FALL IN A FEW MINUTES.

SUNSPOTS ARE COOLER AREAS OF THE SOLAR SURFACE. THEY APPEAR DARK BECAUSE THE REST OF THE SUN IS SO HOT.

A PHOTON TAKES OVER 1 MILLION YEARS TO ESCAPE THE RADIATIVE ZONE.

THE PHOTOSPHERE IS THE SURFACE OF THE SUN. IT RELEASES MOST OF THE SUN'S LIGHT.

THE CHROMOSPHERE IS A LAYER OF GAS ABOVE THE PHOTOSPHERE.

SOME PROMINENCES ARE NOT ERUPTIVE AND CAN LAST FOR A FEW MONTHS.

Black Holes

A black hole is one of the most remarkable objects in the Universe. It is an incredibly dense object, with a gravity so great that nothing can escape from it. Light is the fastest thing in the Universe, but not even light can travel away from a black hole.

To escape the gravitational pull of Earth, a rocket must reach *escape velocity*, a speed of 7 miles (11 km) per second. If Earth were squeezed into a smaller volume, it would become denser. Its gravitational pull would be greater, and thus the escape velocity would increase. If Earth were squeezed to the size of a marble, the escape velocity would increase so much it would be equal to the speed of light. At this point, nothing could escape its gravitational pull, not even light, and Earth would become a black hole.

RADIO LOBES ARE PATCHES OF RADIO WAVES AT THE ENDS OF RADIO JETS.

Active galaxies
Some galaxies contain supermassive black holes. These galaxies are called active galaxies. Often, high-energy jets shoot matter and anti-matter from their cores far into space. Particles in the jets collide and produce radio waves that a radio telescope can detect. The radio waves above have been color-coded by a computer. Only some active galaxies produce radio waves — scientists do not yet know why.

THE TORUS IS OFTEN SURROUNDED BY CLOUDS OF GAS THAT ARE TURNING INTO STARS.

THE TORUS IS A LARGE DISK OF DUST AND GAS SURROUNDING A SUPERMASSIVE BLACK HOLE.

Stellar black holes
A stellar black hole is created when a massive star explodes and collapses from its own gravity. The black hole Cygnus X-1 (1), for example, was created when a massive star in a double star system exploded. The second star (2) is now being pulled to pieces and swallowed by the black hole. As they orbit each other, gas flows from the star onto a disk surrounding the black hole. The gas gets so hot that it gives off X-rays. X-ray telescopes detect this invisible light. The black hole then sucks in the gas.

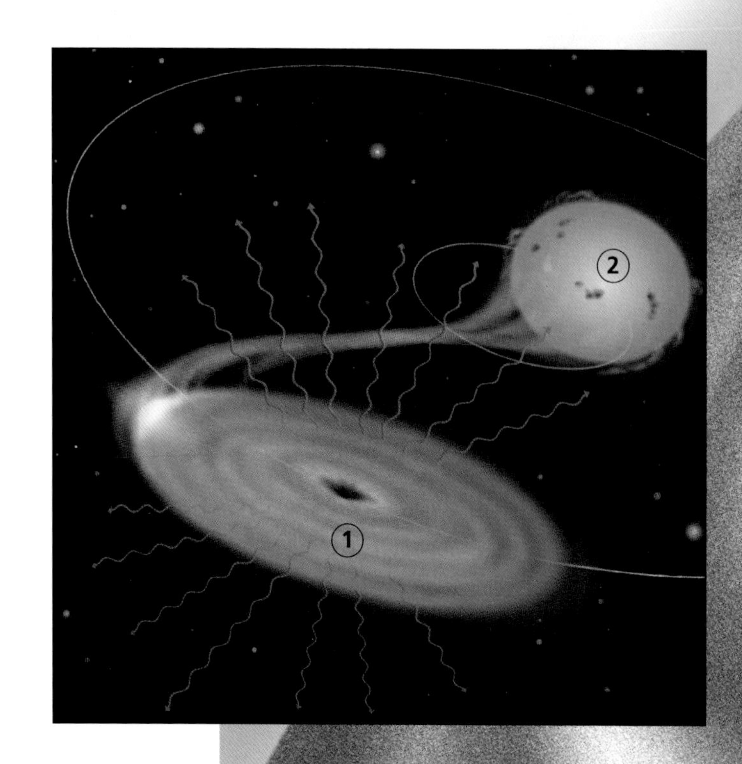

2
1

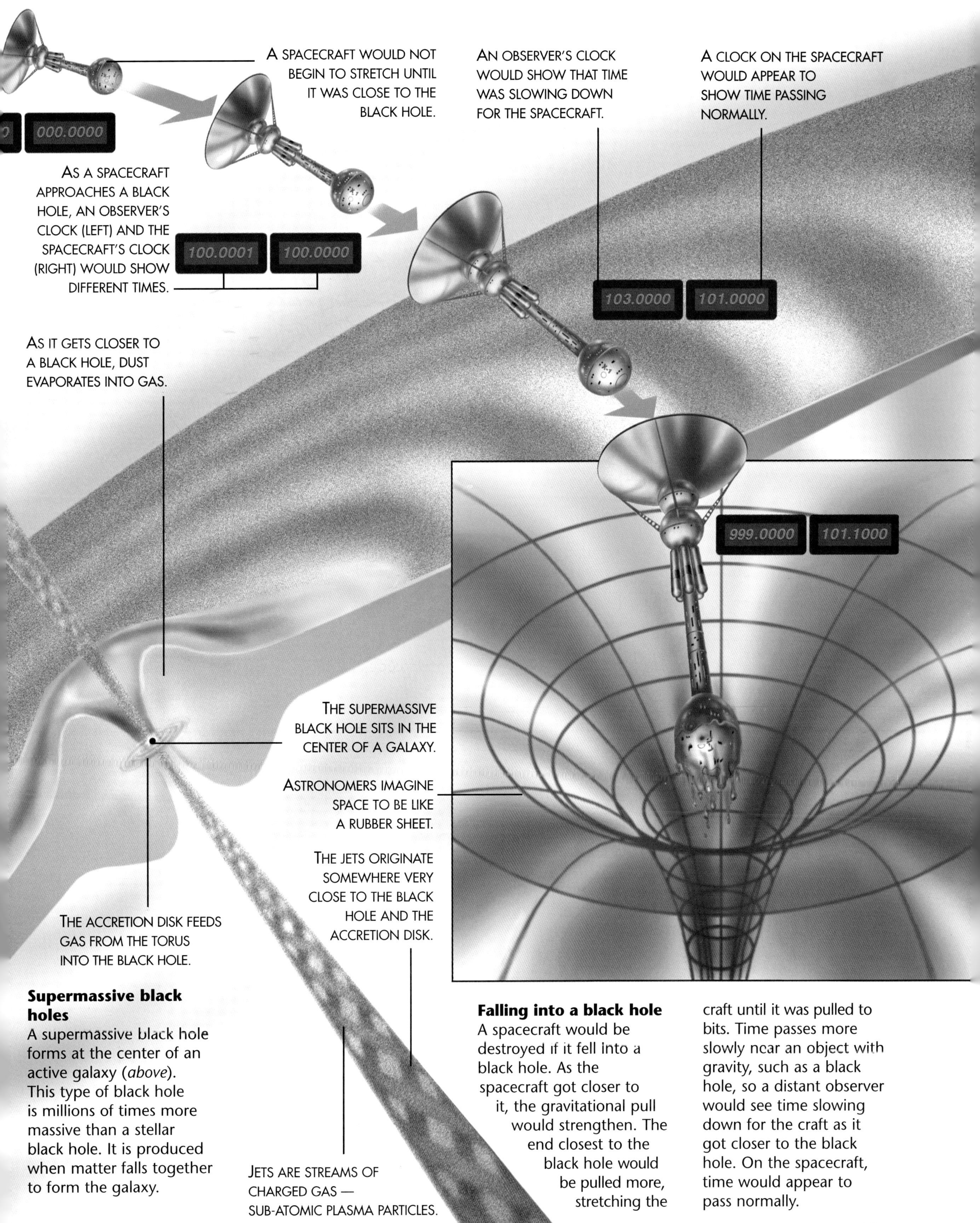
A SPACECRAFT WOULD NOT BEGIN TO STRETCH UNTIL IT WAS CLOSE TO THE BLACK HOLE.
AN OBSERVER'S CLOCK WOULD SHOW THAT TIME WAS SLOWING DOWN FOR THE SPACECRAFT.
A CLOCK ON THE SPACECRAFT WOULD APPEAR TO SHOW TIME PASSING NORMALLY.
000.0000
AS A SPACECRAFT APPROACHES A BLACK HOLE, AN OBSERVER'S CLOCK (LEFT) AND THE SPACECRAFT'S CLOCK (RIGHT) WOULD SHOW DIFFERENT TIMES.
100.0001
100.0000
103.0000
101.0000
AS IT GETS CLOSER TO A BLACK HOLE, DUST EVAPORATES INTO GAS.
999.0000
101.1000
THE SUPERMASSIVE BLACK HOLE SITS IN THE CENTER OF A GALAXY.
ASTRONOMERS IMAGINE SPACE TO BE LIKE A RUBBER SHEET.
THE JETS ORIGINATE SOMEWHERE VERY CLOSE TO THE BLACK HOLE AND THE ACCRETION DISK.
THE ACCRETION DISK FEEDS GAS FROM THE TORUS INTO THE BLACK HOLE.
JETS ARE STREAMS OF CHARGED GAS — SUB-ATOMIC PLASMA PARTICLES.

Supermassive black holes

A supermassive black hole forms at the center of an active galaxy (*above*). This type of black hole is millions of times more massive than a stellar black hole. It is produced when matter falls together to form the galaxy.

Falling into a black hole

A spacecraft would be destroyed if it fell into a black hole. As the spacecraft got closer to it, the gravitational pull would strengthen. The end closest to the black hole would be pulled more, stretching the craft until it was pulled to bits. Time passes more slowly near an object with gravity, such as a black hole, so a distant observer would see time slowing down for the craft as it got closer to the black hole. On the spacecraft, time would appear to pass normally.

Radio Telescopes

For centuries, telescopes have collected light and focused it into images. Light, however, is not the only type of radiation released by objects in space. Many objects also emit X-ray or radio waves. A special radio telescope is needed to collect this radiation.

Radio telescopes are very large dishes. They collect radio waves and focus the waves onto a small detector in the center of the dish. The study of radio waves from space has given astronomers new ideas about galaxies. They have learned, for example, that some active galaxies release powerful X-rays or radio waves (*see page 40*).

Some astronomers also use radio telescopes to search for messages from other life in the Universe. Radio waves might communicate such messages, but astronomers have not yet detected any.

Radio images from space
Radio waves picked up by a radio telescope's detectors are combined into a single image. In this image, different radio intensities are color-coded. To obtain higher quality images, several radio telescopes can be used together. Two or more radio telescopes can observe the same object at the same time. The signals received by each telescope can then be compared and used to form detailed pictures.

The Effelsberg Radio Telescope
The Effelsberg Radio Telescope has a collecting dish with a diameter of 328 feet (100 m). It collects radio waves with this dish and focuses them onto three detectors. The detectors collect radio waves from three slightly different parts of the sky. This speeds up the process of building an image.

THE MAIN DISH IS LIKE THE PRIMARY MIRROR IN A REFLECTING TELESCOPE.
A STRONG FRAMEWORK SUPPORTS THIS RADIO DISH.
WAVEGUIDES CHANNEL RADIO WAVES ONTO PROBES.
PROBES CONVERT RADIO WAVES INTO ELECTRICAL SIGNALS.

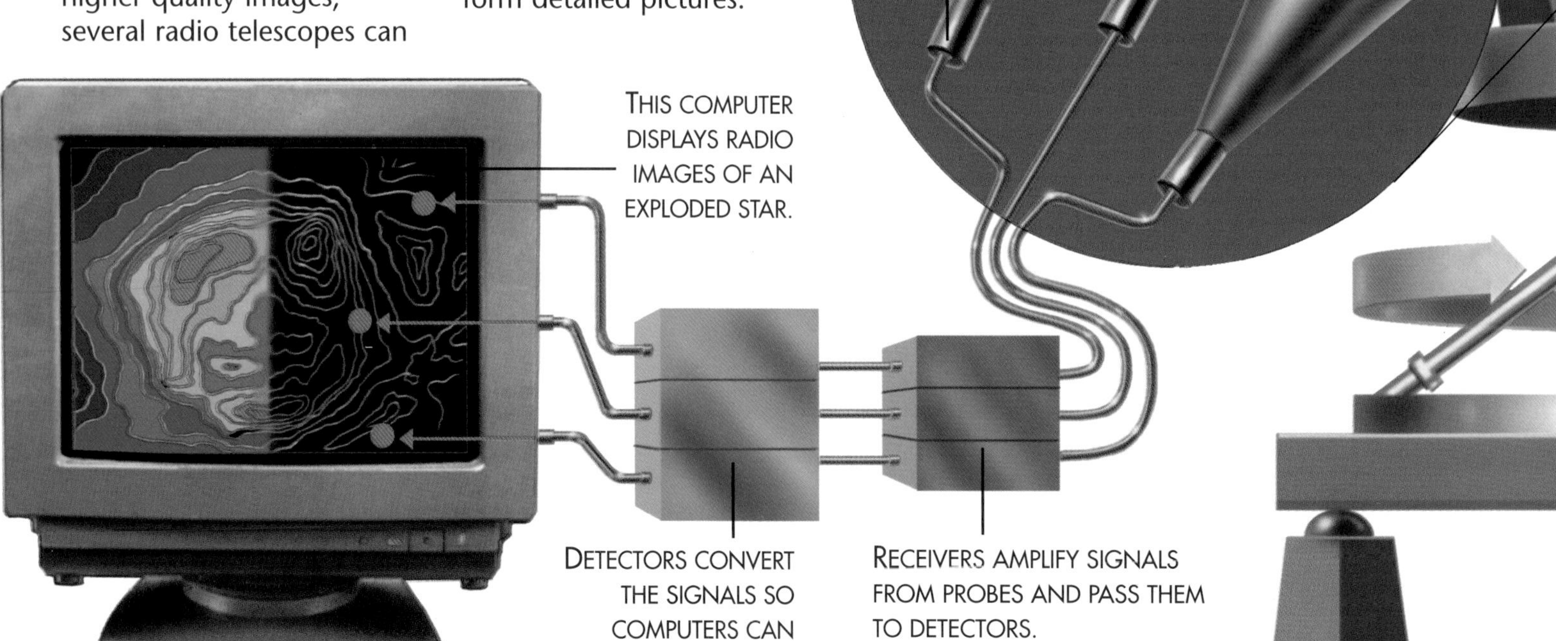
THIS COMPUTER DISPLAYS RADIO IMAGES OF AN EXPLODED STAR.
DETECTORS CONVERT THE SIGNALS SO COMPUTERS CAN READ THEM.
RECEIVERS AMPLIFY SIGNALS FROM PROBES AND PASS THEM TO DETECTORS.

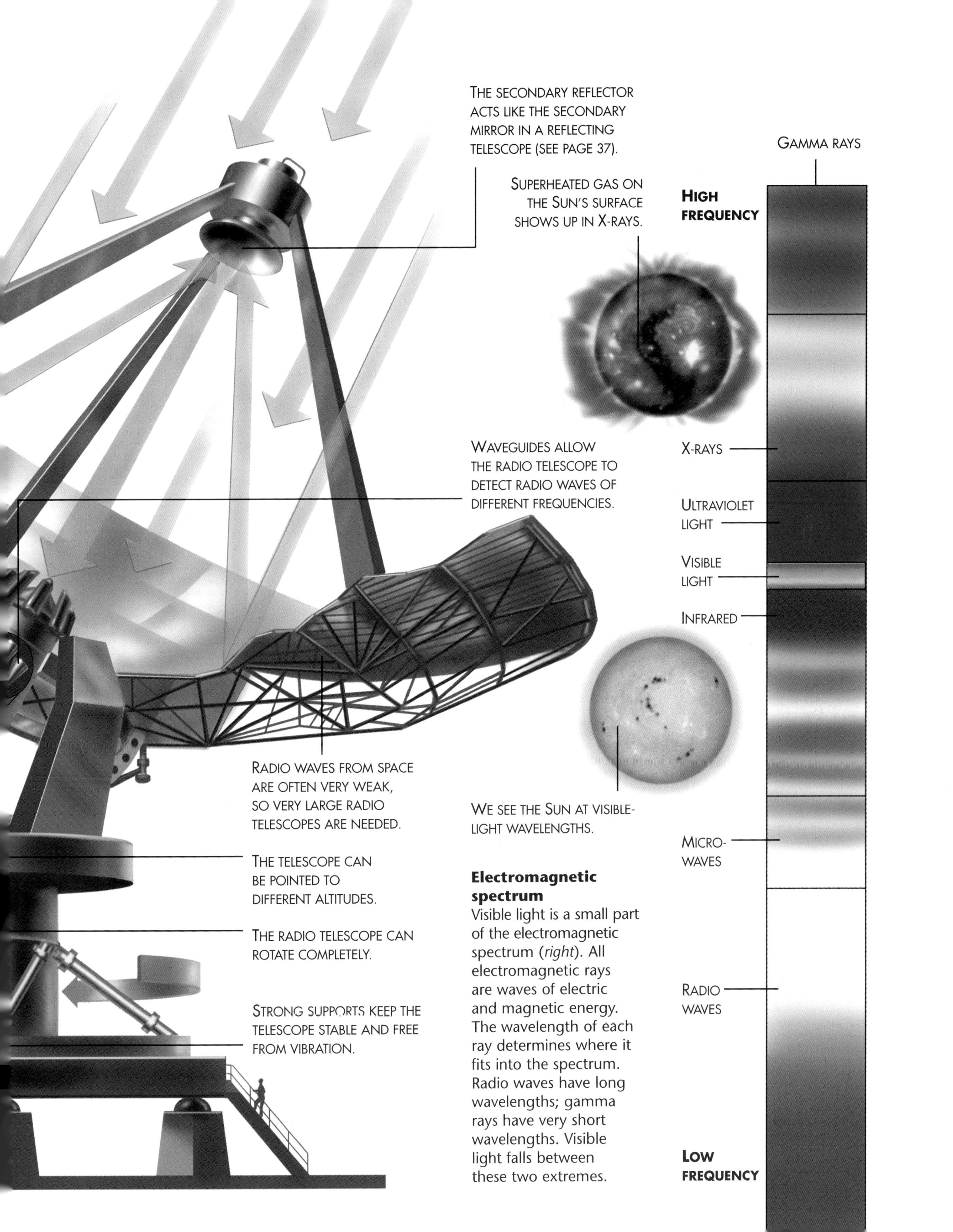
THE SECONDARY REFLECTOR ACTS LIKE THE SECONDARY MIRROR IN A REFLECTING TELESCOPE (SEE PAGE 37).
SUPERHEATED GAS ON THE SUN'S SURFACE SHOWS UP IN X-RAYS.
WAVEGUIDES ALLOW THE RADIO TELESCOPE TO DETECT RADIO WAVES OF DIFFERENT FREQUENCIES.
RADIO WAVES FROM SPACE ARE OFTEN VERY WEAK, SO VERY LARGE RADIO TELESCOPES ARE NEEDED.
WE SEE THE SUN AT VISIBLE-LIGHT WAVELENGTHS.
THE TELESCOPE CAN BE POINTED TO DIFFERENT ALTITUDES.
THE RADIO TELESCOPE CAN ROTATE COMPLETELY.
STRONG SUPPORTS KEEP THE TELESCOPE STABLE AND FREE FROM VIBRATION.
GAMMA RAYS
HIGH FREQUENCY
X-RAYS
ULTRAVIOLET LIGHT
VISIBLE LIGHT
INFRARED
MICRO-WAVES
RADIO WAVES
LOW FREQUENCY

Electromagnetic spectrum
Visible light is a small part of the electromagnetic spectrum (*right*). All electromagnetic rays are waves of electric and magnetic energy. The wavelength of each ray determines where it fits into the spectrum. Radio waves have long wavelengths; gamma rays have very short wavelengths. Visible light falls between these two extremes.

Magnetic Fields

A magnetic field is generated every time an electrically charged object moves. Most planets in the Solar System generate magnetic fields. Earth's magnetic field is produced in the fluid outer core of its center. The heat of the inner core pushes the outer core's fluid up and around in a process called convection (*see page 14*). Because this outer core is made of metal, which can be electrically charged, this convection produces a magnetic field.

The planets Mercury, Venus, and Mars generate magnetic fields in similar ways. Compared to Earth, they have weak magnetic fields. In the center of Jupiter and Saturn, hydrogen gas is compressed until it behaves like a metal, and this generates magnetic fields. Scientists have not yet discovered the magnetism-generating material in Uranus and Neptune.

Auroral display
Tiny particles carrying electrical charges are always streaming away from the Sun. Sometimes these particles get caught in Earth's magnetic field. They are drawn into our atmosphere near the North and South Poles. These charged particles glow and look like streaks of colored light when they hit molecules in the atmosphere. This is called an auroral display.

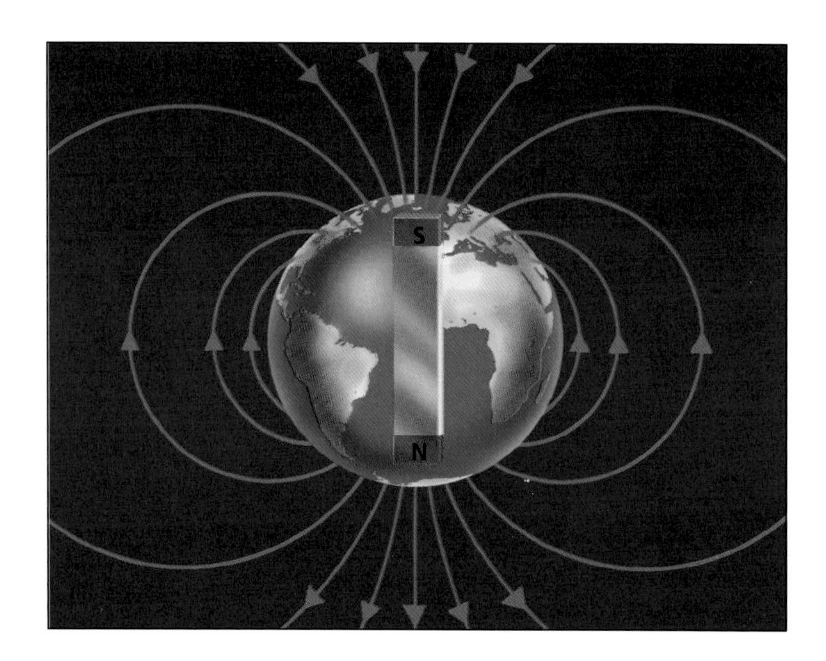
S
N

Magnetic poles
All magnetic fields behave as if a gigantic bar magnet were buried in the center of the planet. Invisible field lines of magnetic force come out through the magnetic North Pole of the planet and come back in through the magnetic South Pole. The magnetic poles are at opposite ends to the geographic poles.

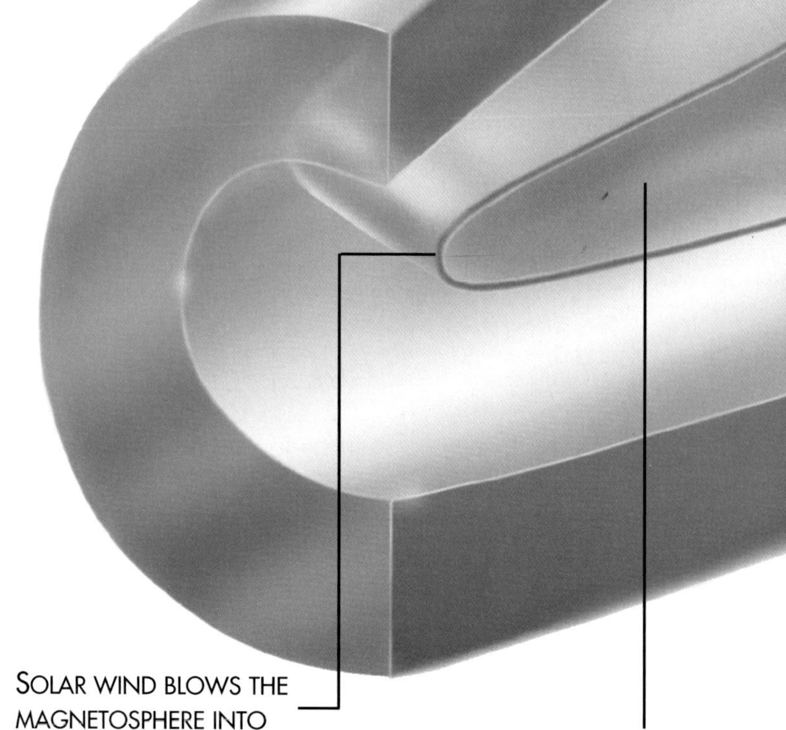
THE MAGNETOSPHERE PROTECTS EARTH FROM THE HARMFUL CHARGED PARTICLES OF THE SOLAR WIND.
MOST PARTICLES FROM THE SUN STREAM PAST EARTH.
SOLAR WIND BLOWS THE MAGNETOSPHERE INTO AN OBLONG SHAPE THAT STRETCHES AWAY FROM THE SUN FOR MILLIONS OF MILES (KM).
SOME PARTICLES ARE CAUGHT AND TRAPPED IN THE PLASMA SHEET.

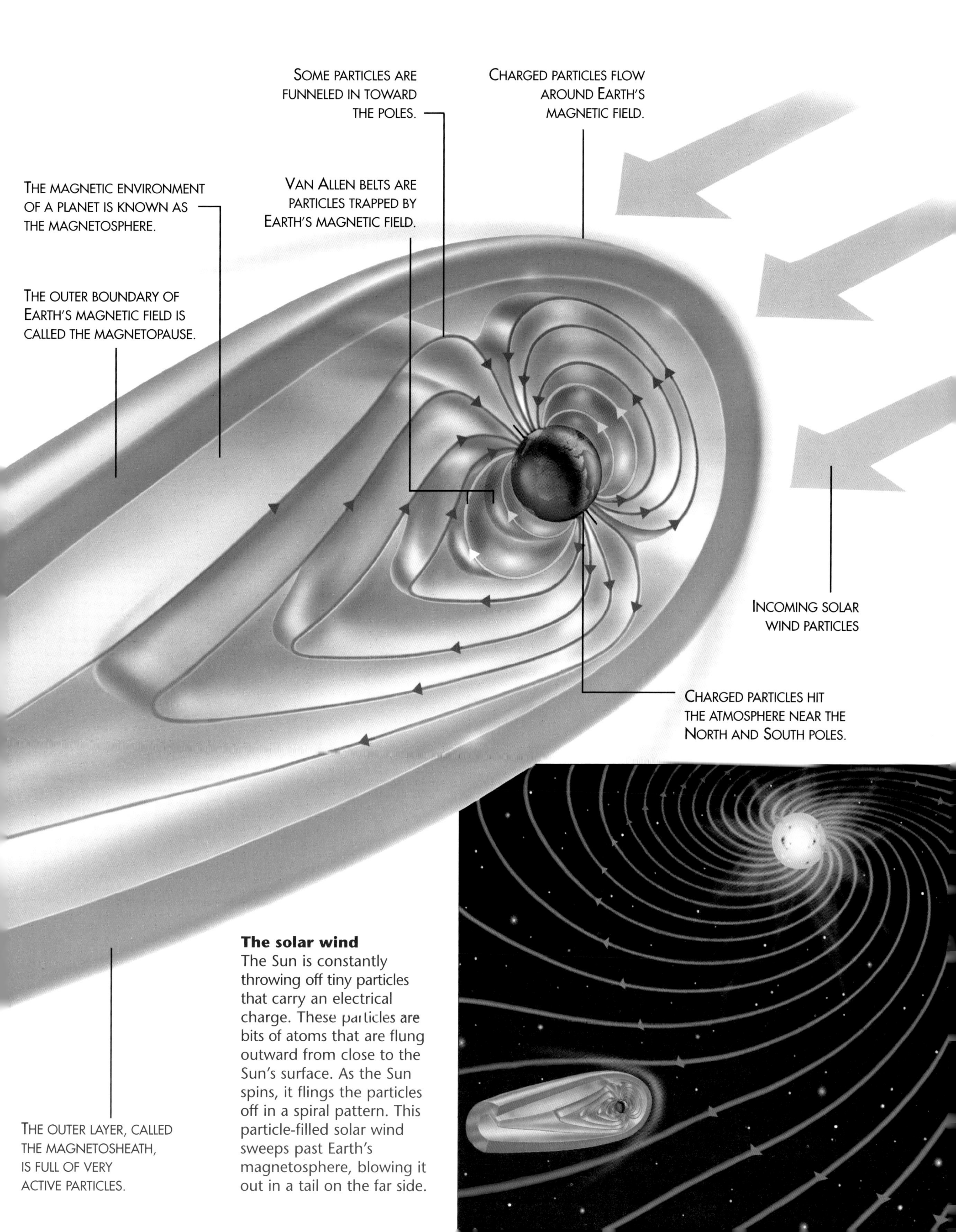
SOME PARTICLES ARE FUNNELED IN TOWARD THE POLES.
CHARGED PARTICLES FLOW AROUND EARTH'S MAGNETIC FIELD.
THE MAGNETIC ENVIRONMENT OF A PLANET IS KNOWN AS THE MAGNETOSPHERE.
VAN ALLEN BELTS ARE PARTICLES TRAPPED BY EARTH'S MAGNETIC FIELD.
THE OUTER BOUNDARY OF EARTH'S MAGNETIC FIELD IS CALLED THE MAGNETOPAUSE.
INCOMING SOLAR WIND PARTICLES
CHARGED PARTICLES HIT THE ATMOSPHERE NEAR THE NORTH AND SOUTH POLES.
THE OUTER LAYER, CALLED THE MAGNETOSHEATH, IS FULL OF VERY ACTIVE PARTICLES.

The solar wind
The Sun is constantly throwing off tiny particles that carry an electrical charge. These particles are bits of atoms that are flung outward from close to the Sun's surface. As the Sun spins, it flings the particles off in a spiral pattern. This particle-filled solar wind sweeps past Earth's magnetosphere, blowing it out in a tail on the far side.

Glossary

accretion: the process of growing by building up or collecting material.

asteroids: small, planetary bodies made of rock or metal. Many asteroids in the Solar System orbit the Sun in the asteroid belt between Mars and Jupiter.

astronomer: a scientist who studies the stars, planets, and other occurrences in the Universe.

atmosphere: the layer of gases that surrounds planets and stars.

atom: the smallest unit of a chemical element that can exist alone.

billion: the number represented by 1 followed by nine zeros — 1,000,000,000. In the British system, it is called "a thousand million."

celestial: in the starry sky; part of outer space.

convection: the circular movement of fluid of different temperatures, caused by the different thicknesses of the fluid and the effects of gravity on the fluid; the transferring of heat through this movement.

elliptical: shaped like an oval.

equator: an imaginary line that divides a planet into two equal hemispheres.

filament: a gigantic group of superclusters; a string.

galaxy: a very large group of stars and planets.

generate: to create or produce.

geologist: a scientist who studies Earth's layers of soil and rock.

gravity: the force that pulls smaller objects toward larger objects. Objects that have a greater mass, or weight, have a stronger force of gravity.

jets: pairs of high-energy streams of electrically charged particles, or plasma, that shoot out thousands of light-years from a black hole center.

light-year: a unit of measurement equal to the distance light travels in a vacuum in one year. This distance is 5.9 trillion miles (9.5 trillion km).

magnetic field: an area around a magnet within which the magnet's power attracts metals.

magnetic pole: one of the two points where a planet's magnetic pull is the strongest.

module: a unit that can operate independently but is also a part of the total structure of a space vehicle.

nuclear fusion: the joining of two atomic nuclei to form a new, heavier nucleus. This process creates much energy that is released as light and heat in stars.

nucleus: the center of an atom. This center is positively charged and contains smaller particles called neutrons and protons.

orbit: (v) to travel around or revolve around; (n) the path an object follows as it travels around another object in space.

quark: one of the tiny particles composing the parts of an atom.

redshift: the stretching of a galaxy's light, as seen from Earth. This stretching is caused by the expansion of space and the effect of Earth's gravity on wavelengths.

reflecting telescope: an instrument that makes faraway objects seem larger by using mirrors, rather than lenses, to collect and focus light.

refracting telescope: an instrument that makes faraway objects seem larger by using lenses, rather than mirrors, to collect and focus light.

satellite: a spacecraft that orbits Earth. Satellites are used for scientific study or for communications, usually by television or telephone.

solar system: a star and all the celestial bodies held by its gravity.

solar wind: a stream of charged particles from the Sun that flows through the Solar System.

star: a huge ball of very hot gas that gives off heat and light produced by nuclear fusion.

supercluster: a large group of galaxies.

trillion: the number represented by 1 followed by twelve zeros — 1,000,000,000,000. In the British system, this number is called "a billion."

velocity: the speed of an action.

More Books to Read

Astronauts: Training for Space. Countdown to Space (series). Michael D. Cole (Enslow)

Beyond the Stars. James Passmore (Nature Publishing House)

Close Encounters: Exploring the Universe with the Hubble Space Telescope. Elaine Scott (Disney Press)

Crafts for Kids Who Are Wild about Outer Space. Crafts for Kids Who Are Wild (series). Kathy Ross (Millbrook Press)

Deep Space Astronomy. Gregory Vogt (Twenty-First Century Books)

DK Guide to Space. Peter Bond (Dorling Kindersley)

Galaxies. Window on the Universe (series). Robert Estalella (Barrons Juveniles)

Modern Astronomy. Isaac Asimov's New Library of the Universe (series). Isaac Asimov, Greg Walz-Chojnacki, and Frank Reddy (Gareth Stevens)

Our Planetary System. Isaac Asimov's New Library of the Universe (series). Isaac Asimov, Greg Walz-Chojnacki, and Francis Reddy (Gareth Stevens)

Our Star: The Sun. Window on the Universe (series). Robert Estalella and Marcel Socias (Barrons Juveniles)

Videos

The Apollo Project. (International Video Corporation)
Astronomers. (MPI Home Video)
Destination: Space. (National Geographic)
The Planets. (A & E Entertainment)
Super Structures of the World: International Space Station. (Unapix)

Web Sites

Future Astronauts of America
www.faahomepage.org/

NASA for Kids
kids.msfc.nasa.gov/

NASA's Observatorium
observe.ivv.nasa.gov/

The Nine Planets
www.seds.org/nineplanets/nineplanets/

The Observatory
www.exploratorium.edu/observatory/index.html

Mysteries of Deep Space
www.pbs.org/deepspace/

Sky and Telescope — Tips
www.skypub.com/tips/tips.shtml

Windows to the Universe
msgc.engin.umich.edu/

Some web sites stay current longer than others. For further web sites, use your search engines to locate the following topics: *astronauts, black holes, northern lights,* Pathfinder, *planets, Solar System,* and *stars.*

Index